AF785688

Reliabilities of Consecutive-k Systems

Network Theory and Applications

Volume 4

Managing Editors:

Ding-Zhu Du, *University of Minnesota, U.S.A.*

and

Cauligi Raghavendra, *University of Southern California, U.S.A.*

The titles published in this series are listed at the end of this volume.

Reliabilities of Consecutive-k Systems

by

Gerard J. Chang

Department of Applied Mathematics,
National Chiao Tung University,
Hsinchu, Taiwan, R.O.C.

Lirong Cui

Reliability and Safety Research Center,
China Aerospace Corporation,
P.R. China

and

Frank K. Hwang

Department of Applied Mathematics,
National Chiao Tung University,
Hsinchu, Taiwan, R.O.C.

KLUWER ACADEMIC PUBLISHERS
DORDRECHT / BOSTON / LONDON

A C.I.P. Catalogue record for this book is available from the Library of Congress.

ISBN-13:978-1-4613-7972-0 e-ISBN-13:978-1-4613-0273-5
DOI: 10.1007/978-1-4613-0273-5

Published by Kluwer Academic Publishers,
P.O. Box 17, 3300 AA Dordrecht, The Netherlands.

Sold and distributed in North, Central and South America
by Kluwer Academic Publishers,
101 Philip Drive, Norwell, MA 02061, U.S.A.

In all other countries, sold and distributed
by Kluwer Academic Publishers,
P.O. Box 322, 3300 AH Dordrecht, The Netherlands.

Printed on acid-free paper

All Rights Reserved
© 2000 Kluwer Academic Publishers
Softcover reprint of the hardcover 1st edition 2000

No part of the material protected by this copyright notice may be reproduced or
utilized in any form or by any means, electronic or mechanical,
including photocopying, recording or by any information storage and
retrieval system, without written permission from the copyright owner.

Contents

List of Figures

Preface

The consecutive-k system was first studied around 1980, and it soon became a very popular subject. The reasons were many-folded, including:

1. The system is simple and natural. So most people can understand it and many can do some analysis. Yet it can grow in many directions and there is no lack of new topics.

2. The system is simple enough to become a prototype for demonstrating various ideas related to reliability. For example, the interesting concept of component importance works best with the consecutive-k system.

3. The system is supported by many applications.

Twenty years have gone and hundreds of papers have been published on the subject. This seems to be a good time for retrospect and to sort the scattered material into a book. Besides providing our own perspective, the book will also serve as an easy reference to the numerous ramifications of the subject. It is hoped that a summary of work in the current period will become the seed of future break-through.

GERARD J. CHANG, LIRONG CUI, FRANK K. HWANG

Chapter 1

INTRODUCTION

1.1 THE CONSECUTIVE SYSTEM AND GENERALIZATIONS

The *consecutive system*, also known as the *consecutive-k-out-of-n:F system* in the literature, is a system of n components arranged in a line such that the system fails if and only if some k consecutive components fail. The system was first studied by Kontoleon [131] under the name of the *r-successive-out-of-n:F system*, but only an enumerating algorithm was given. The paper by Chiang and Niu [61] motivated the system by some real applications, showed that the reliability of the system could be computed efficiently by recursive equations, and christened the current name. This paper laid down the foundation and popularized the study of the consecutive system. In this book, we will use the term the *consecutive-k-out-of-n system*, the *consecutive-k system* or the *consecutive system* interchangeably depending on whether there is a need to bring out the parameters k and n (F is always suppressed and understood).

The consecutive system as a research topic got another boost from the paper of Derman, Lieberman and Ross [68]. They made two fundamental contributions which had lasting impact in the field. The first is that they extended the original linear systems to circular systems: n components arranged into a cycle. The second, and the more important contribution, is that they introduced the optimal design aspect into the problem. In particular, their conjecture of how n components with distinct reliabilities should be assigned to the n positions in a linear system to maximize the reliability of a consecutive-2 system has attracted attention from discrete mathematicians and operations researchers.

Shanthikumar [235] and Hwang [101] first applied time complexity analysis on proposed algorithms for computing the system reliabilities; thus exposing the theoretical computer science aspect of the consecutive system problem. On the other hand, Chao and Lin [52] and Fu [77] started the study of the asymptotic behavior of the system when n goes to infinity, ensuring that the probability theory plays an important role of its own.

There are two basic assumptions for the consecutive system:

(i) **Binary system.** The components and the systems all have two states—*working* or *failed.*

(ii) **IND model.** The states of the components are independent. If, furthermore, they are identical, we call it the *IID model.*

Both assumptions have been generalized. Dependency of component states has been introduced into the probability model, and binary state has been extended to multi-state. But the results on these generalized models are limited. By and large, the literature on the consecutive system adopts the above two assumptions.

There are other variations which differ in appearance but not in substance. For example, we can exchange the role of working and failed components and define a *consecutive-k:G system.* Namely, the system works if and only if some k consecutive components all work. However, the reliability of the G system is simply the unreliability of the F system computed by switching the component reliabilities and its unreliabilities.

The basic model takes a snapshot view of the system, i.e., it computes the system reliability at a certain instant when component i has the probability p_i of working. We may also study the dynamic behavior of the system by asking the availability of the system at time t given that the probability that component i works from time 0 up to time t is $p_i(t)$. However, the system availability at time t is simply the system reliability computed by substituting p_i with $p_i(t)$.

Although the above two variations do not lead to new problems in computing reliability, they do lead to new problems in other aspects. For example, to maximize the reliability of a G system is equivalent to minimize the reliability of an F system which is different from to maximize the reliability of an F system. Similarly, the dynamic system induces new problems like mean time to failure and mean number of failed components at system failures.

The consecutive system has also been extended to some more general systems. By treating the system failure as the failure to send a flow from the first component to the last component through the linear system, it is natural to take components as nodes each of which has links to the next k nodes. Thus if k consecutive nodes all fail, then the node immediately preceding them cannot relay the flow forward and the system fails. By bringing out the links into the open, we may assume that links can also work or fail. This is called the *network system*. This network system can be further generalized into the graph system by removing the restriction that the underlying graph is a line or a cycle.

The requirement that k consecutive components must fail to fail the system has also been extended to some k out of m consecutive components must fail. In general, we call any m consecutive components an *m-window* and call a window *bad* if it contains k or more failed components. Then we set the criterion for a system to fail in terms of bad window. For examples, a system failure can be induced by a single bad window, or by all windows being bad, or in the general case by b or more bad windows. Conceptually, one can also specify a system failure by b or more consecutive bad windows. Note that the consecutive-k system can either be interpreted as a window system which fails on one bad k-window where a k-window is bad if it contains only failed components, or as a window system which fails on k or more bad 1-windows where a 1-window is a single component.

Finally, the book also covers some topics like high-dimensional consecutive systems and multi-failure systems which have strong application need and whose theory is under development, as well as a few other topics.

1.2 THE PROBLEMS AND THE METHODOLOGIES

The book considers two general types of problems: computing the system reliability (or some other system statistics) and maximizing the system reliability through optimally assigning components to positions in the system.

There are two general approaches to compute the system reliability for the IND probability models. The *recursive equation approach* which was successfully pursued from the very beginning by Chiang and Niu [61] and Derman, Lieberman and Ross [68]. The *embedded Markov chain approach* was first proposed by Griffith and Govindarajula [96], Griffith [95], Chao and Lin [52], and perfected by Fu [79]. A variation, the *transfer matrix method*, was recently advocated by Hwang and Wright [117].

The recursive equation approach sets up a recursive equation to compute the system reliability. Often there are various ways to set up recursive equations, and one has to analyze their computational complexities to make a choice. Occasionally, a recursive equation is not self-evident and one has to introduce exogenous variables or equations to form a system of equations whose solution yields the desired recursive equation.

A recursive equation with w parameters can be transformed into a difference equation with w variables. If the difference equation has a single variable and is of degree at most four, then a closed-form solution exists which can be transformed back as a solution for the recursive equation. A difference equation with more than one variable is sometimes solvable under more stringent conditions.

For the embedded Markov chain approach, it is important to recognize that the elements of the Markov chain are not the components, which are assumed to be independent, but the windows (consisting of consecutive components). Again, there may be more than one Markov chains embedded in the computation of system reliability. We are looking for the Markov chain which can be defined with a small number of states such that the probability transition matrix has reasonable size and the computation can be performed efficiently. For example, for the consecutive-k system, a k-window has 2^k original states. Fu defined a Markov chain which has only $k+1$ states representing the number of consecutive failed components at the end of a window.

Note that a Markov chain needs an initial state. Hence the embedded Markov chain approach does not apply to circular systems.

The transfer matrix approach is similar to the embedded Markov chain approach in appearance as the transfer matrix operates like a probability transition matrix. Furthermore, there is also an ordering of the state spaces and the transfer matrix deals with the transition of the states in one space to the states of the succeeding space. Two states in consecutive state spaces are defined over disjoint components, hence the à priori probabilities of the two states are independent, and can be separated from the transition probability. By choosing a proper way to embed these à priori probabilities into the transition matrices, the circular system reliability can be computed by multiplying these transition matrices around and taking the trace of the product.

For the IID probability model, there is the additional combinatorial approach which counts the number of ways the system fails given some parameters and then sums up the ranges of parameters attached with the proper probabilities. One efficient way of doing this is using the generating function.

The general optimal allocation problem is to allocate $m \geq n$ components with nonidentical reliabilities to the n positions in a system, where a position can be allocated any number of components to maximize the system reliability. It is assumed that each component can go to any position (the components are functionally equivalent), and a position works if any of its allocated components works. There are also two different versions of this general problem depending on information available:

(i) **Sequential.** Components are allocated one at a time. As soon as a component is inserted into a position, we know its state.

(ii) **Nonadaptive.** Components are allocated simultaneously.

Clearly, the system reliability of the sequential design will be higher since more information is available.

Since the general problem is very difficult, in most cases we do not have a globally optimal algorithm. Sometimes a heuristic, which can be a locally optimal algorithm, is used.

The literature has not dealt with the optimal allocation in its full generality. The two special cases which have been studied are:

(i) $m = n$ under the IND model.

(ii) $m > n$ under the IID model.

For case (ii), the problem also varies according to the initial conditions. The two prevalent models are either starting from scratch, or starting from having one component at each position.

1.3 NOTATION

In this book, we adopt the following notation.

$\Pr(e)$	probability of event e.
$\Pr(e_1 \mid e_2)$	probability of event e_1 under the condition that event e_2 occurs.
$E(X)$	expectation of a random variable X.
$E(X \mid e)$	expectation of a random variable X under the condition that event e occurs.
$\mathrm{Trace}(M)$	trace of a matrix M.
n	number of components in the system.
p_i	probability that component i functions.
q_i	$1 - p_i$.
$L(1,n)$ or L	consecutive-k-out-of-n linear system, k and all p_i are assumed understood.
$C(1,n)$ or C	consecutive-k-out-of-n circular system, k and all p_i are assumed understood.
$R_x(1,n)$ or $R(x)$	reliability of the system $x(1,n)$ where $x \in \{L, C\}$, k and all p_i are assumed understood.
$\overline{R}_x(1,n)$	$1 - R_x(1,n)$, unreliability of the system $x(1,n)$.
$R_x(n)$	$R_x(1,n)$ when all $p_i = p$ and all $q_i = q$.
$\overline{R}_x(n)$	$1 - R_x(n)$, unreliability of the system $x(n)$.

s_i	working state of component i.
$\bar{s}_i$	failure state of component i.
$x(i,j)$	x system containing components $i, i+1, \ldots, j$.
$R_x(i,j)$	reliability of the subsystem $x(i,j)$.
$\overline{R}_x(i,j)$	$1 - R_x(i,j)$, unreliability of the subsystem $x(i,j)$.
$S_x(i,j)$	set of all component states including a work $x(i,j)$.
$\overline{S}_x(i,j)$	set of all component states including a failed $x(i,j)$.
$N_x(d,n,k)$	number of ways that n nodes, including exactly d failed nodes, ordered in a line (if $x = L$) or cycle (if $x = C$) contains no k consecutive failed nodes.

Chapter 2

COMPUTATION OF RELIABILITY

We discuss various methods to compute the reliability of a consecutive system and compare their time complexities.

2.1 RECURSIVE EQUATIONS

Chiang and Niu [61] gave the first solution of the reliability of a linear consecutive-k system under the IID model. This solution is in the form of recursive equations. Many other recursive equations have been proposed since with the objective to improve the computational complexity. While most of the recursive equations are stated in terms of reliability, their arguments are always derived from the more basic structural relation, a relation based on the states. This was first pointed out by Rushdi [214], also Chan, Chan and Lin [37]. We will state the recursive equations in its structure form.

Shanthikumar [235] and Hwang [101] independently gave the following solution to the linear system.

THEOREM 2.1.1 $S_L(1,n-1) = S_L(1,n) \cup S_L(1,n-k-1) s_{n-k} \bar{s}_{n-k+1} \cdots \bar{s}_n.$

Proof. If $L(1,n)$ works, then of course $L(1,n-1)$ works. The only case that $L(1,n-1)$ works but $L(1,n)$ doesn't is when the last $k+1$ nodes have the states as specified. ■

COROLLARY 2.1.2 $R_L(1,n) = R_L(1,n-1) - R_L(1,n-k-1) p_{n-k} \prod_{j=1}^{k} q_{n-k+j}.$

Proof. The two events stated in Theorem 2.1.1 are disjoint. ■

Note that

$$\{t_i \equiv \prod_{j=1}^{k} q_{i+j} : i = 0, 1, \ldots, n-k\}$$

can be computed in $O(n)$ time by noting the relation

$$t_{i+1} = t_i q_{i+k+1}/q_{i+1}.$$

Then each $R_L(1,h)$, computed in the order $h = 1, 2, \ldots, n$, is the difference of two numbers and each can be computed in $O(1)$ time. It follows that $R_L(1,n)$, which can be computed only after $R_L(1,1), \ldots, R_L(1,n-1)$ are all computed, takes $O(n)$ time to compute. This is the fastest time complexity for both the IID and the IND models using recursive equations. To compare, the recursive equations given by Chiang and Niu [61] take $O(kn^2)$ time to compute, and the following recursive equation given by Hwang [101] takes $O(kn)$ time.

THEOREM 2.1.3 $S_L(1,n) = \bigcup_{n-k+1}^{n} S_L(1, i-1) s_i \bar{s}_{i+1} \ldots \bar{s}_n.$

Proof. Let i denote the last working node. Then the system works if and only if $i \geq n-k+1$ and the subsystem consisting of the first $i-1$ nodes work. ■

COROLLARY 2.1.4 $R_L(1,n) = \sum_{n-k+1}^{n} \left(R_L(1, i-1) p_i \prod_{j=i+1}^{n} q_j \right).$

For the IID model, Chang, Chen and Hwang [43] extended Corollary 2.1.2 to

THEOREM 2.1.5 *Setting* $R_L(-1) = 1/p$ *and* $R_L(n) = 0$ *for* $n \leq -2$. *Then*

(i) $R_L(n) = R_L(i) R_L(n-i) - pq^k \sum_{s=k+1-i}^{k} R_L(n-i-s)$ *for* $1 \leq i < k$.

(ii) $R_L(n) = R_L(i) R_L(n-i) - p^2 q^k \sum_{s=2}^{k} R_L(n-i-s) \sum_{r=k+2-s}^{k} q^{r+s-k-2} R_L(i-r)$ *for* $k \leq i \leq \lceil n/2 \rceil$.

Proof. Let S_n, S_i and S_{n-i} denote the original system and its two subsystems consisting of the first i components and the last $n-i$ components respectively. Then S_n fails while both S_i and S_{n-i} work if and only if there exists a set F of at least k consecutive failed components, including components i and $i+1$, satisfying the following conditions:

Case (i). $s-1$ components in F are in S_{n-i} and are followed by a working component; at least $k-s+1$ components in F are in S_i.

Case (ii). $s-1$ components in F are in S_{n-i} and are followed by a working component; $r-1$ components in F are in S_i and are preceded by a working component.

The negative terms in (i) and (ii) simply sum up all such choices. The reason to set $R(-1) = 1/p$ is to take care of the case all components in S_i or S_{n-i} are in F. Then there is no room (and no need) for the working component preceding or following these failed components. ∎

Note that Corollary 2.1.2 is the special case $i = 1$. For constant i, the first recursive equation can compute $R_L(n)$ in $O(n)$ time. The motivation of Theorem 2.1.5 is really not for computation, but for providing a property which will be useful in proving component importance (Sec. 3.3). Also note that Theorem 2.1.5 can be easily stated for the IND model.

Derman, Lieberman and Ross [68] first gave a solution for the reliability of the circular system under IID model, also in the recursive equation form. Their argument implies the following result. Note that indices are taken modulo n for circular systems.

THEOREM 2.1.6 $S_C(1,n) = \bigcup_{\substack{1\leq i\leq k \\ n-k+i\leq j\leq n}} s_j\bar{s}_{j+1}\ldots\bar{s}_n\bar{s}_1\ldots\bar{s}_{i-1}s_iS_L(i+1,j-1).$

Proof. The case of $n < k$ is trivial. Suppose $n \geq k$. Then $S_C(1,n)$ must contain at least a working node. Let i denote the working node with the minimum index and $j(\geq i)$ the working node with the maximum index. Then there must be at most $k-1$ nodes (all failed) between node j and node i (from j to i). The range given in the union covers all such cases. Furthermore, the subsystem (linear) from node $i+1$ to node $j-1$ must work. ∎

COROLLARY 2.1.7 $R_C(1,n) = \sum_{\substack{1\leq i\leq k \\ n-k+i\leq j\leq n}} p_j\left(\prod_{h=j+1}^{n+i-1} q_h\right)p_iR_L(i+1,j-1).$

COROLLARY 2.1.8 $R_C(n) = p^2\sum_{i=0}^{k-1}(i+1)q^iR_L(n-i-2).$

$R_C(n)$ given in Corollary 2.1.8 can be computed in $O(kn)$ time. Again

$$\{t_{j,i} = \prod_{h=j+1}^{n+i-1} q_h : 1 \leq i \leq k, n-k+i \leq j \leq n\}$$

can be separately computed in $O(k^2)$ time. Hwang [101] observed that there are $O(k^2)$ R_L terms in the sum each requiring $O(n)$ time to compute. Hence $R_C(1,n)$ takes $O(k^2n)$ time to compute. Wu and Chen

[256] observed that the time complexity can be improved to $O(kn)$ since $R_L(i,i), R_L(i,i+1), \ldots, R_L(i,n)$ can be computed together in $O(n)$ time for each $i = 1, 2, \ldots, k$.

Wu and Chen [255] introduced a trick of using fictitious nodes to make the $O(kn)$ time more explicit. Define $L(1, n+i)$ to be a system where $p_1 = p_2 = \ldots = p_i = p_{n+1} = p_{n+2} = \ldots = p_{n+i} = 0$.

THEOREM 2.1.9 $\overline{S}_C(1,n) = \overline{S}_L(1,n) \cup \bigcup_{i=1}^{k-1} \left(\overline{S}_L(1,n+i) \setminus \overline{S}_L(1,n+i-1)\right)$.

Proof. Let T_i denote the event $\overline{s}_i\overline{s}_{i+1}\ldots\overline{s}_{i+k-1}$. Then

$$\begin{aligned}\overline{S}_C(1,n) &= T_1 \cup \overline{T}_1T_2 \cup \overline{T}_1\overline{T}_2T_3 \cup \ldots \cup \overline{T}_1\ldots\overline{T}_{n-1}T_n \\ &= \overline{S}_L(1,n) \cup \overline{T}_1\ldots\overline{T}_{n-k+1}T_{n-k+2} \cup \ldots \cup \overline{T}_1\ldots\overline{T}_{n-1}T_n.\end{aligned}$$

But $\overline{T}_1\ldots\overline{T}_{n-k+i}T_{n-k+i+1}$ is equal to the ith term in the union in Theorem 2.1.9. ■

By setting $p_j = 0$ for the fictitious nodes $j = i+1, \ldots, i+k-1$, we have

COROLLARY 2.1.10 $\overline{R}_C(1,n) = \overline{R}_L(1,n) + \sum_{i=1}^{k-1}\prod_{j=1}^{i} q_j\left(\overline{R}_L(1,n+i) - \overline{R}_L(1,n+i-1)\right)$.

Since there are $O(k)$ terms of R_L, it takes $O(kn)$ time to compute.

There are other recursive equations also leading to an $O(kn)$-time algorithm. Antonopoulou and Papastavridis [7] gave the following recursive equation.

THEOREM 2.1.11 $S_C(1,n) = s_nS_L(1,n-1) \cup \left(\overline{s}_nS_C(1,n-1) \setminus \bigcup_{i=1}^{k} s_{n-k+i-1}\,\overline{s}_{n-k+i}\ldots\overline{s}_n\overline{s}_1\ldots\overline{s}_{i-1}s_i)S_L(i+1, n-k+i-2)\right)$.

Proof. If node n is working, then $C(1,n)$ works if and only if $L(1,n-1)$ works. If node n is failed, then $C(1,n)$ works if and only if $C(1,n-1)$ works except in one case that the latter works but not the former. That is, when exactly k consecutive nodes, including node n, are failed. ■

COROLLARY 2.1.12 $R_C(1,n) = p_nR_L(1,n-1) + q_nR_C(1,n-1) - \sum_{i=1}^{k} p_{n-k+i-1}\left(\prod_{j=n-k+i}^{n+i-1} q_j\right) p_iR_L(i+1, n-k+i-2)$.

Since the equation in Corollary 2.1.12 has only k R_L terms, it takes $O(kn)$ time to compute.

There exists a faster recursive equation for the IID model. Lambiris and Papastarvridis [152] stated the following result without proof. Du and Hwang [73] gave a proof and asked for a direct proof. We give a direct proof here.

THEOREM 2.1.13 *$R_C(n) = R_C(n-1) - R_C(n-k-1)pq^k$ for $n \geq 2k+1$.*

Proof. Let i denote the working node with the minimum index. Let $C(n-1)$ be the cycle determined from $C(n)$ by deleting node $i+1$. Then $C(n)$ works if and only if $C(n-1)$ works except when nodes $i+1, \ldots, i+k$ all failed but node $i+k+1$ working, then $C(n)$ is failed but $C(n-1)$ working. The restriction to $n \geq 2k+1$ is needed because if $i = k$, then node $i+k+1$ is node 1 which cannot be working by the definition of i. ■

Note that the formula in Corollary 2.1.2 for IID model is

$$R_L(n) = R_L(n-1) - R_L(n-k-1)pq^k,$$

which has the same form as Theorem 2.1.13. The solving of such a recursive equation into closed form reduces to the solving of the algebraic equation

$$x^n = x^{n-1} - pq^k x^{n-k-1} \quad \text{or} \quad x^{k+1} = x^k - pq^k.$$

Since $x = q$ is a root, this equation can be solved for $2 \leq k \leq 4$. Let $r_1, r_2, \ldots, r_k$ denote the roots. In the case of distinct roots, a general solution is

$$R(n) = \sum_{i=1}^{k} \alpha_i r_i,$$

where α_i are determined by the initial conditions. For example, for $k = 2$, the initial conditions for the linear system are

$$R_L(2) = 1 - q^2 \quad \text{and} \quad R_L(3) = 1 - q^3 - 2pq^2;$$

and those for the circular system are

$$R_C(2) = 1 - q^2 \quad \text{and} \quad R_C(3) = 1 - q^3 - 3pq^2.$$

Tong [245] gave a closed form formula for $n \leq 2k+1$.

THEOREM 2.1.14 *For $n \le 2k+1$,*

$$R_C(1,n) = \begin{cases} 1 - \prod_{i=1}^{n} q_i, & \text{if } k = n; \\ 1 - \sum_{j=1}^{n} \left(p_j \prod_{i=1}^{k} q_{j+i} \right) - \prod_{i=1}^{n} q_i, & \text{if } k \le n-1. \end{cases}$$

Proof. For $n \le 2k+1$, the cycle can obtain at most one run of failed nodes with length at least k. When not all nodes are failed, then such a run is preceded by a working node. Let C_j denote the minimal cutset preceded by working node j. Then the C_j's, $1 \le j \le n$, are disjoint, and together they contain all cutsets. Hence Theorem 2.1.14. ■

Zuo and Kuo [270] obtained the following

COROLLARY 2.1.15 *For $n \le 2k$,*

$$R_L(1,n) = 1 - \sum_{i=1}^{n-k+1} \left(p_{i+k} \prod_{j=i}^{i+k-1} q_j \right),$$

where $p_{n+1} = 1$.

2.2 THE MATRIX APPROACH

Griffith and Govindarajula [96] and Griffith [95] introduced Markov chain into the problem in study. However, their approach does not lead to an efficient algorithm. Chao and Lin [52] used the Markov chain in a different way. Although their goal was not for obtaining an algorithm, their theory has lent base to the development of a very efficient algorithm.

Viewing the system as a dynamic process starting from node 1 and adding one node at a time, it is convenient to talk about the system state S_v at node v. Chao and Lin defined S_v as the aggregate of states of nodes $v-k+1, v-k+2, \ldots, v$. Therefore S_{v+1} depends only on S_v, and $\{S_v\}$ forms a Markov chain. Fu [79] reduced the number of states, while preserving the Markov chain property, by defining S_v to be in state i, $i \in \{0, 1, \ldots, k\}$, if the last i nodes including node v are all failed. Note that state k is the failed state while all other states are sub-states of the working state. Fu and Hu [81] gave the following $(k+1) \times (k+1)$ transition probability matrix for S_v

$$\Lambda_v = \begin{pmatrix} p_v & q_v & & & \\ p_v & & q_v & & \\ \vdots & & & \ddots & \\ p_v & & & & q_v \\ 0 & & & & 1 \end{pmatrix}$$

where row i and column i represent the state i, $i = 0, 1, \ldots, k$. Let π_0 denote the $1 \times (k+1)$ row vector $(1, 0, \ldots, 0)$ and let U_0 denote the $(k+1) \times 1$ column vector of all 1 except the last bit is 0. Then

$$s(1,n) = \pi_0 \prod_{v=1}^{n} \Lambda_v U_0,$$

where π_0 specifies that the initial system state is 0, and U_0 sums up the probability of the working states at node n.

While Fu and Hu [81] gave no time complexity analysis, Koutras [140] later restated this method and claimed it computes $R_L(1,n)$ in $O(kn)$ time. Multiplying two order-$(k+1)$ matrices in general takes $O(k^3)$ time. The sparseness of Λ_v reduces the computation of each cell to be constant, but there are still $(k+1)^2$ cells to fill. Therefore each multiplication takes $O(k^2)$ time, and we have $n-2$ multiplications of matrices. Therefore it takes $O(k^2 n)$ time to compute $R_L(1,n)$.

Koutras [140] discussed the IID case (where $\Lambda_v = \Lambda$) without giving a different time complexity analysis. But it is well known (for example Hwang and Wright [117] used it in their analysis) that M^n can be obtained by computing $M^2, M^4, \ldots, M^{2^{\lfloor \log_2 n \rfloor}}$ only. Therefore, it takes $O(k^3 \log n)$ time to compute $R_L(n)$.

Since a Markov chain must have a starting point, this method doesn't work for the circular system.

Hwang and Wright [117] gave a different Markov chain method (called the transfer matrix method) which is better described by the structural function. For easier presentation, assume k divides n. Partition the n nodes into n/k sets of k each, where the ith set consists of nodes $(i-1)k+1, (i-1)k+2, \ldots, ik$. A state of set i is simply a vector of states of its k elements.

Let $S_i = \{S_{iu}\}$ denote the state space of set i Then $|S_i| = 2^k$. Let M_i denote the matrix whose rows are indexed by members of S_i, and columns by members of S_{i+1}. Cell $(u,v) = 1$ if $S_{iu} \cup S_{(i+1)v}$ does not contain k consecutive failed nodes, and $= 0$ otherwise. For the circular system, let $S_{n/k+1} = S_1$. Then

$$\text{Trace}\left(\prod_{i=1}^{n/k} M_i\right)$$

contains the number of working cases out of the 2^n total cases (a case is a complete specification of states for all nodes).

To compute reliability, $\Pr(S_{(i+1)v})$ has to be brought in. Hwang and Wright [117] described a clever way by substituting each entry 1 in cell

(u, v) of M_i with $\sqrt{\Pr(S_{iu})\Pr(S_{(i+1)v})}$, since each S_{iu} appears once in M_i and once in M_{i-1}. For a linear system, then the corresponding term is

$$\prod_{i=1}^{n/k-1} M_i,$$

and $M_1(u, v) = 1$ needs to be replaced by $\Pr(S_{1u})\sqrt{\Pr(S_{2v})}$ and $M_{n/k-1} = 1$ by $\sqrt{\Pr(S_{n/k-1,u})\Pr(S_{n/k,v})}$.

It takes $O(2^{3k})$ time to do each multiplication; there are $O(n/k)$ multiplications for the IND model and $O(\log(n/k))$ multiplications for the IID model.

Hwang and Wright [117] also found a way to cut down the multiplication time from $O(2^{3k})$ to $O(k^6)$. For a working state S_{iu}, let ℓ denote the index of the first working node and $r(\geq \ell)$ the last. For each S_i, delete the unique failed state and regroup the working states according to (ℓ, r). Then the size of $|S_i|$ is cut down from $O(2^k)$ to $O(k^2)$, while the ability to determine whether $S_{iu} \cup S_{(i+1)v}$ contains k consecutive failed nodes is preserved since it depends only on r of S_{iu} and ℓ of $S_{(i+1)v}$. Note that

$$\Pr(S_{i,(\ell,r)}) = \left(\prod_{j=1}^{\ell-1} q_{(i-1)k+j}\right) p_{(i-1)k+\ell} p_{(i-1)k+r} \left(\prod_{j=1}^{k-1} q_{(i-1)k+r+j}\right),$$

but without $p_{(i-1)k+r}$ if $\ell = r$. Therefore the time to compute $R(n)$ for the IND model is $O(k^5 n)$.

For the IID model, $O(k^6)$ can be further reduced to $O(k^3)$. Define

$$h_{\ell r} = \Pr(S_{(\ell,r)}) = \begin{cases} p^2 q^{\ell-1+k-r}, & \text{if } 0 \leq \ell < r \leq k-1; \\ p q^{\ell-1+k-r}, & \text{if } 0 \leq \ell = r \leq k-1; \\ 0, & \text{if } 0 \leq r < \ell \leq k-1. \end{cases}$$

Let $H = (h_{\ell r})$ be a $k \times k$ matrix, and let $I_H = (I_{\ell r})$ be another $k \times k$ matrix obtained from H by setting

$$I_{\ell r} = \begin{cases} 0, & \text{if } h_{\ell r} = 0; \\ 1, & \text{if } h_{\ell r} > 0. \end{cases}$$

Hwang and Wright [117] showed that

$$\text{Trace}\Big((HI)^j\Big) = \text{Trace}(M^j) \quad \text{for } j = 0, 1, \ldots.$$

H and I can be constructed in $O(k^2)$ time, and HI in $O(k^3)$ time. Multiplication of two order $O(k)$ matrices takes $O(k^3)$ times. Hence the time to compute $R(n)$ is $O(k^3 \log(n/k))$.

Example 2.2.1 Suppose $k = 3$. Then

$$H = \begin{pmatrix} p^2 & p^2q & pq^2 \\ p^2q & pq^2 & 0 \\ pq^2 & 0 & 0 \end{pmatrix}, \quad I = \begin{pmatrix} 1 & 1 & 1 \\ 1 & 1 & 0 \\ 1 & 0 & 0 \end{pmatrix}, \quad HI = \begin{pmatrix} p & pq & p^2 \\ pq & pq & p^2q \\ pq^2 & pq^2 & pq^2 \end{pmatrix}.$$

$R(3) = \text{Trace}(HI) = p + pq + pq^2 = 1 - q^3$ is certainly correct. Let $(a_{3j}, a_{3j-1}, \ldots, a_j)$ denote the polynomial $\sum_{i=j}^{3j} a_i p^i q^{3j-i}$. Then, for $j = 2$,

$$(HI)^j = \begin{pmatrix} (1,5,10,6,1) & (1,5,9,4,0) & (1,4,6,0,0) \\ (0,1,4,7,2) & (0,1,4,6,1) & (0,1,3,4,0) \\ (0,0,1,4,3) & (0,0,0,3,4) & (0,0,1,2,1) \end{pmatrix}.$$

$R(6) = \text{Trace}\Big((HI)^2\Big) = p^6 + 6p^5q + 15p^4q^2 + 14p^3q^3 + 3p^2q^4$ is easily checked to be correct. For $j = 3$,

$$\begin{aligned} R(9) &= \text{Trace}\Big((HI)^3\Big) \\ &= (1,8,28,50,45,16,1) + (0,1,7,20,26,13,1) + (0,0,1,5,10,7,1) \\ &= p^9 + 9p^8q + 36p^7q^2 + 75p^6q^3 + 81p^5q^4 + 36p^4q^5 + 3p^3q^6 \end{aligned}$$

can again be verified to be correct.

The case that k does not divide n can be similarly handled except the last set of nodes has a different cardinality than k. This may affect two M matrices, but would not affect the total time complexity. Hwang and Wright also stated that a similar result holds for the linear system.

Thus we see that the time complexity of both Markov chain methods is slightly inferior to the best recurrence equation methods under the IND model, but much superior to those under the IID model.

2.3 THE COMBINATORIAL APPROACH

The combinatorial approach works only for the IID model. Let $N_L(d, n, k)$ denote the number of ways that n nodes, including exactly d failed nodes, ordered in a line contains no k consecutive failed nodes; and let $N_C(d, n, k)$ be the cycle version. Then

$$R_x(n) = \sum_{d \geq 0} N_x(d, n, k) q^d p^{n-d} \quad \text{for } x \in \{L, C\}. \tag{2.3.1}$$

The combinatorial approach focuses on the solving of $N_x(d,n,k)$.

$$N_L(d,n,2) = \binom{n-d+1}{d} \quad \text{and} \quad N_C(d,n,2) = \frac{n}{n-d}\binom{n-d}{d}$$

are given by the Kaplansky's theorems [128] motivated by the solution to the 'problème des ménages', see also Hwang and Yao [121] and Chen and Koh [58]. Derman, Lieberman and Ross [68] proved

$$N_L(d,n,3) = \sum_{i\geq 0}\binom{n-d+1}{i}\binom{n-d+1-i}{d-2i}.$$

More generally, they proved

THEOREM 2.3.1 *For $k \geq 2$,*

$$N_L(d,n,k) = \sum_{i\geq 0}\binom{n-d+1}{i} N_L(d-(k-1)i, n-ki, k-1).$$

Proof. View a line of n nodes as obtained by inserting the d failed nodes into the $n-d+1$ spaces partitioned by the $n-d$ working nodes. Then each space can contain at most $k-1$ failed nodes. Remove every such set of $k-1$ consecutive failed nodes along with an adjacent working node. Note that by proper coordination no working node needs be removed twice. The remaining $n-ki$ nodes, including $d-(k-1)i$ failed ones, have at most $k-2$ failed nodes in a space. On the other hand, every such line of $n-ki$ nodes can be incremented to a line of n nodes by inserting i subsequences of $k-1$ failed nodes preceded or succeeded by a working node into i of the $n-d+1$ spaces partitioned by the $n-d$ working nodes. The one-one mapping is established which proves Theorem 2.3.1. ■

Bolinger [27] gave a different recursive equation.

THEOREM 2.3.2 *For $k \geq 1$,*

$$N_L(d,n,k) = \begin{cases} \binom{n}{d}, & \text{if } d < k; \\ 0, & \text{if } n = d \geq k; \\ \sum_{i=0}^{k-1} N_L(d-i, n-1-i, k), & \text{if } n > d \geq k. \end{cases} \tag{2.3.2}$$

Proof. The two initial conditions are easily verified. Suppose $i+1$ is the first working node. For the line to work, necessarily, $0 \leq i \leq k-1$.

Then the line works if and only if the remaining $n-1-i$ nodes, including $d-i$ failed ones, is a working line too. ■

Lambiris and Papastavridis [152] and Hwang [103], independently obtained a closed form for $N_L(d,n,k)$.

THEOREM 2.3.3 *For $n > d \geq k \geq 1$,*

$$N_L(d,n,k) = \sum_{i\geq 0}(-1)^i \binom{n-d+1}{i}\binom{n-ki}{n-d}. \tag{2.3.3}$$

Proof. We may prove the theorem by simply showing that the closed form is a solution of (2.3.2).

$$\begin{aligned}
&\sum_{j=0}^{k-1} N_L(d-j, n-1-j, k) \\
&= \sum_{j=0}^{k-1}\sum_{i\geq 0}(-1)^i\binom{n-d}{i}\binom{n-1-j-ki}{n-1-d} \\
&= \sum_{i\geq 0}(-1)^i\binom{n-d}{i}\sum_{j=0}^{k-1}\binom{n-1-j-ki}{n-1-d} \\
&= \sum_{i\geq 0}(-1)^i\binom{n-d}{i}\left(\binom{n-ki}{n-d}-\binom{n-k-ki}{n-d}\right) \\
&\quad \text{(Graham et al [93], (5.10) in Page 160)} \\
&= \sum_{i\geq 0}(-1)^i\left(\binom{n-d}{i}+\binom{n-d}{i-1}\right)\binom{n-ki}{n-d} \\
&= \sum_{i\geq 0}(-1)^i\binom{n-d+1}{i}\binom{n-ki}{n-d} \\
&= N_L(d, npp, k).
\end{aligned}$$

■

Cui, Hawkes and Jalali [65] gave another solution which is less attractive in computing time:

THEOREM 2.3.4 *For $k \geq 3$,*

$$N_L(d,n,k) = \sum \binom{n-d+1}{t_1}\prod_{s=2}^{k-1}\binom{t_{s-1}}{t_s},$$

where the summation runs over $t_1 \geq t_2 \geq \ldots \geq t_{s-1} \geq t_s \geq \ldots \geq t_{k-1} \geq t_k = 0$ and $\sum_{s=1}^{k-1} t_s = d$.

Proof. Let t_s represent the number of intervals between good components into which at least s consecutive faulty items are placed. Then, clearly, if d faulty items are placed so that no k are consecutive, we must have

$$t_1 \geq t_2 \geq \ldots \geq t_{s-1} \geq t_s \geq \ldots \geq t_{k-1} \geq t_k = 0 \text{ and } \sum_{s=1}^{k-1} t_s = d.$$

Then, at each stage we must choose t_s of the t_{s-1} intervals into which we place one extra faulty component. The theorem then follows ■

THEOREM 2.3.5

$$\begin{aligned} R_L(n) &= \sum_{i\geq 0}(-1)^i p^{i-1} q^{ki}\left(\binom{n-ki+1}{i} - q\binom{n-ki}{i}\right) \\ &= \sum_{i\geq 0}(-pq^k)^i\binom{n-ki}{i} - q^k\sum_{i\geq 0}(-pq^k)^i\binom{n-ki-k}{i}. \end{aligned}$$

Proof. Substituting (2.3.3) into (2.3.1),

$$\begin{aligned} R_L(n) &= \sum_{d\geq 0} q^d p^{n-d}\sum_{i\geq 0}(-1)^i\binom{n-d+1}{i}\binom{n-ki}{n-d} \\ &= \sum_{i\geq 0}(-1)^i\frac{(n-ki)!}{i!\big(n-(k+1)i+1\big)!} \\ &\quad \sum_{d\geq 0} q^d p^{n-d}(n-d+1)\binom{n-(k+1)i+1}{d-ki} \\ &= \sum_{i\geq 0}(-1)^i\frac{(n-ki)!}{i!\big(n-(k+1)i+1\big)!}q^{ki}p^{i-1} \\ &\quad \sum_{j\geq 0} q^j p^{n-(k+1)i+1-j}(n-ki+1-j)\binom{n-(k+1)i+1}{j} \\ &= \sum_{i\geq 0}(-1)^i\frac{(n-ki)!q^{ki}p^{i-1}}{i!\big(n-(k+1)i+1\big)!}\Big(n-ki+1-\big(n-(k+1)i+1\big)q\Big) \\ &= \sum_{i\geq 0}(-1)^i p^{i-1}q^{ki}\left(\binom{n-ki+1}{i} - q\binom{n-ki}{i}\right) \\ &= \sum_{i\geq 0}(-1)^i p^{i-1}q^{ki}\left(\binom{n-ki+1}{i} - (1-p)\binom{n-ki}{i}\right) \end{aligned}$$

$$= \sum_{i\geq 0}(-pq^k)^i\binom{n-ki}{i} - q^k\sum_{i\geq 0}(-pq^k)^i\binom{n-ki-k}{i}.$$

■

Papastavridis and Hadjichristos [195] obtained a variation.

COROLLARY 2.3.6 $R_L(n) = \sum_{m=0}^{(k+2)n}(-p)^m \sum_{i=m-n}^{m}\left(\binom{n-ki}{i}\binom{ki}{m-i} - \binom{n-ki-k}{i}\binom{ki+k}{m-i}\right).$

Proof. By expanding $q^x = (1-p)^x = \sum_{j=0}^{x}\binom{x}{j}(-p)^j$ and setting $m = i+j$ in changing variables. ■

Hwang and Yao [121] gave $N_C(d,n,k)$ in terms of $N_L(d,n,k)$ as follows:

THEOREM 2.3.7 $N_C(d,n,k) = \frac{n}{n-d}N_L(d,n-1,k).$

Proof. Let $L_{d,n}$ ($C_{d,n}$) denote the set of working n-lines (n-cycles) containing exactly d failed components. Let C_i' denote the subset of $C_{d,n}$ in which the ith component works. Then there exists a one-to-one mapping from C_i' to L_{n-1}. Thus

$$|C_i'| = |L_{d,n-1}| = N_L(d,n-1,k).$$

Each $C_{d,n}$ has $n-d$ working components and can generate a set of $n-d$ C_i'. Counting the sum of $|C_i'|$ in two ways, we obtain

$$|C_{d,n}|(n-d) = \sum_{i=1}^{n}|C_i'| = n \cdot N_L(d,n-1,k).$$

Therefore

$$N_C(d,n,k) = |C_{d,n}| = \frac{n}{n-d}N_L(d,n-1,k).$$

■

Substituting $R_L(n)$ into Corollary 2.1.10 yields a closed form (in double summation) solution of $R_C(n)$. Both Lambris and Papastavridis [152] and Hwang [103] managed to reduce it to one sum as in

THEOREM 2.3.8

$$R_C(n) = -q^n + \sum_{i\geq 0}(-pq^k)^i\binom{n-ki}{i} + k\sum_{i\geq 0}(-pq^k)^{i+1}\binom{n-ki-k-1}{i}.$$

Proof. We give the proof by Goulden [92] using generating functions. The coefficient of $x^i y^j$ in a bivariate formal power series $g(x, y)$ is denoted by $[x^i y^j]g(x, y)$. Let T be the set of sequences of m successes (S) and $n - m$ failures (F) containing no occurrences of k circularly consecutive failures. Each elements of T is repeated m times (once for each occurrence of S) in the n circular rotations of sequences of length n in $\{S(\epsilon \cup F \cup F^2 \cup \ldots \cup F^{k-1})\}^m$, where ϵ is an empty string. Thus

$$N_C(d, n, k) = [x^d y^{n-d}]G(x, y)$$

where

$$\begin{aligned} G(x,y) &= \sum_{m\geq 1} \frac{1}{m}\left(y\left(\frac{1-x^k}{1-x}\right)\right)^m + \sum_{i=1}^{k-1} \frac{x^i}{i} \\ &= \log\left(1 - y(1-x^k)(1-x)^{-1}\right)^{-1} + \sum_{i=1}^{k-1} \frac{x^i}{i} \end{aligned}$$

and the second summation accounts for sequences composed entirely of failures. Use the operator identity $n[z^n] = [z^n]z\frac{d}{dz}$, we obtain

$$\begin{aligned} R_C(n) &= [z^n]z\frac{d}{dz}\left(\log(1-qz) + \log(1 - z + pq^k z^{k+1})^{-1}\right) \\ &= -q^n + [z^n]\left(\frac{1 - kpq^k z^{k+1}}{1 - z + pq^k z^{k+1}} - 1\right) \\ &= -q^n + [z^n](1 - kpq^k z^{k+1})(1-z)^{-1}\left(1 + pq^k z^{k+1}(1-z)^{-1}\right)^{-1} \\ &= -q^n + [z^n]\sum_{i\geq 0}(-pq^k)^i\left(z^{(k+1)i} - kpq^k z^{(k+1)(i+1)}\right)(1-z)^{-(i+1)}. \end{aligned}$$

∎

Hwang [103] gave explicit solutions (without summation) of $R_L(n)$ and $R_C(n)$ for $k = 2$.

THEOREM 2.3.9 *For $k = 2$, let $\rho = q/p$. Then*

$$R_L(n) = \frac{p^n}{2^n\sqrt{1+4\rho}}\left((1+\sqrt{1+4\rho})^{n+1} - (1-\sqrt{1+4\rho})^{n+1}\right), \quad (2.3.4)$$

$$R_C(n) = \frac{p^n}{2^n}\left((1+\sqrt{1+4\rho})^n + (1-\sqrt{1+4\rho})^n\right). \quad (2.3.5)$$

Proof. For $k = 2$

$$\begin{aligned}
R_L(n) &= \sum_{d=0}^{\lceil n/2 \rceil} q^d p^{n-d} N_L(d, n, 2) \\
&= \sum_{d=0}^{\lceil n/2 \rceil} q^d p^{n-d} \binom{n+1-d}{d} \\
&= p^n \sum_{d=0}^{\lceil n/2 \rceil} \binom{n+1-d}{d} \rho^d \\
&= \text{RHS of (2.3.4)} \quad \text{(Graham et al [93], (5.74) in Page 204)} \\
R_C(n) &= \sum_{d=0}^{\lfloor n/2 \rfloor} q^d p^{n-d} N_C(d, n, 2) \\
&= p^n \sum_{d=0}^{\lfloor n/2 \rfloor} \binom{n-d}{d} \frac{n}{n-d} \rho^d \\
&= \text{RHS of (2.3.5)} \quad \text{(Graham et al [93], (5.75) in Page 204)}
\end{aligned}$$

■

2.4 BOUNDS AND APPROXIMATIONS

Since bounds and approximations must be simple to be useful, they are usually given for the IID model; although the ideas behind them can work also for the IND model.

Feller [75], page 325, used partial fraction expansion of generating function to obtain an approximation of $R_L(n)$:

$$R_L(n) \sim \frac{1 - qs}{(k + 1 - ks)ps^{n+1}},$$

where s is the unique positive root of x in

$$1 - px(1 + qx + \ldots + q^{k-1}x^{k-1}) = 0.$$

Feller commented that the approximation is surprisingly good even for very small n, and improves rapidly with n. For example, for $k = 2$, the error rate is 0.0163 for $n = 2$, 0.0080 for $n = 3$, 0.0016 for $n = 4$ and 0.0005 for $n = 5$.

Many bounds are closely related to the notion of cutset. A *cutset* is a set of nodes whose simultaneous failures causes the system failure. A cutset is *minimal* if no proper subset is a cutset. Note that the system fails if and only if the nodes in a certain minimal cutset are all failed. Since the minimal cutsets are not disjoint, the inclusion-exclusion principle needs to be used to obtain $\overline{R}_L(n)$. This approach of computing $\overline{R}_L(n)$ has not been pursued since it is enumerative in nature and since many efficient algorithms to compute $\overline{R}_L(n)$ are known. However, one can truncate the terms in the inclusion-exclusion formula to obtain bounds. For example

$$\overline{R}_L(n) = \Pr\Big(\bigcup_i \text{ minimal cutset } i\Big) \le \sum_i \Pr(\text{minimal cutset } i).$$

A similar bound can be argued for $R_L(n)$.

$$\begin{aligned} R_L(n) &= \Pr\Big(\bigcap_i (\text{nonoccurence of minimal cutset } i\Big) \\ &\ge \prod_i \Pr(\text{nonoccurence of minimal cutset } i). \qquad (2.4.1) \end{aligned}$$

Chiang and Niu [61] first applied the minimal cutset notion to the linear system.

THEOREM 2.4.1 $(1-q^k)^{n-k+1} \le R_L(n) \le (1-q^k)^{\lfloor n/k \rfloor}$.

Proof. The set of all minimal cutsets is

$$\{\overline{S}_L(1,k), \overline{S}_L(2,k+1), \ldots, \overline{S}_L(n-k+1,n)\}.$$

The probability of occurrence of each minimal cutset is q^k, and the probability of nonoccurence is $1-q^k$. So the lower bound follows from (2.4.1).

Let C denote a set of disjoint minimal cutsets. Then

$$\begin{aligned} R_L(n) &\le \Pr\Big(\bigcap_i (\text{nonoccurence of minimal cutset } i \in C)\Big) \\ &= \prod_i \Pr(\text{nonoccurence of minimal cutset } i \in C). \end{aligned}$$

The upper bound is obtained by choosing

$$C = \{S_L(1,k), S_L(k+1,2k), \ldots, S_L((\lfloor n/k \rfloor - 1)k+1, \lfloor n/k \rfloor k)\}.$$

■

Salvia [219] gave the easier-to-compute (but less accurate) version.

COROLLARY 2.4.2 $1-(n-k+1)q^k \le R_L(n) \le 1-(n-k+1)p^{n-k}q^k$.

Derman, Lieberman and Ross [68] extended Theorem 2.4.1 to the IND model and also to the circular system. They pointed out that better bounds can be obtained by working out the reliability of some small system and apply the bounds to the remaining subsystem. For example

LEMMA 2.4.3 *For* $1 \leq m \leq n-1$,

$$R_L(1,n) \geq R_L(1,m) \prod_{i=m-k+2}^{n-k+1} \left(1 - \prod_{j=i}^{i+k-1} q_j\right)$$

and

$$R_C(1,n) \geq R_C(1,m) \prod_{i=m-k+2}^{n} \left(1 - \prod_{j=i}^{i+k-1} q_j\right).$$

Chen and Hwang [55] proposed a bound by setting $m = 3k$ in the linear model.

Derman, Lieberman and Ross [68] proposed an upper bound for $R_C(n)$. First, a lemma.

LEMMA 2.4.4 *If* Y *be a nonnegative random variable, then*

$$\Pr(Y > 0) \geq E^2(Y)/E(Y^2).$$

Proof. Note that

$$\begin{aligned} E(Y^2) &= E(Y^2|Y>0)\Pr(Y>0) \\ &\geq E^2(Y|Y>0)\Pr(Y>0) \quad \text{by the Cauchy-Schwarz inequality} \\ &= E^2(Y)/\Pr(Y>0). \end{aligned}$$

The lemma then follows ∎

Let $\overline{G}_i$ denote the event $\overline{S}(i-k+1,\ldots,i)$, and $\overline{G}$ denote the number of $\overline{G}_i$.

COROLLARY 2.4.5 $R_x(1,n) \leq 1 - E^2(\overline{G})/E(\overline{G}^2)$ *for* $x \in \{L, C\}$.

Note that

$$E(\overline{G}) = \begin{cases} \sum_{i=1}^{n-k+1} \prod_{j=i}^{i+k-1} q_j, & \text{for the linear system;} \\ \sum_{i=1}^{n} \prod_{j=i}^{i+k-1} q_j, & \text{for the circular system,} \end{cases}$$

with $q_{n+j} = q_j$. $E(\overline{G}^2)$ can be similarly computed, though messier. The circular case has a simple form for the IID model. First

LEMMA 2.4.6 *Let Y denote the number of events occurred among the events $A_1, A_2, \ldots, A_r$. If*

$$\Pr(A_i|Y=j) = j/r \text{ for all } j \le r,$$

then

$$E(Y|A_i) = E(Y^2)/E(Y).$$

Proof. Note that

$$\begin{aligned}
E(Y|A_i) &= \sum_{j=0}^{r} j\Pr(Y=j|A_i) \\
&= \sum_{j=0}^{r} j\Pr(A_i|Y=j)\Pr(Y=j)/\Pr(A_i) \\
&= \sum_{j=0}^{r} j^2\Pr(Y=j)/(r\Pr(A_i)) \\
&= E(Y^2)/(r\Pr(A_i)); \\
\Pr(A_i) &= \sum_{j=0}^{r} \Pr(A_i|Y=j)\Pr(Y=j) \\
&= \sum_{j=0}^{r} j\Pr(Y=j)/r \\
&= E(Y)/r.
\end{aligned}$$

The lemma then follows. ∎

THEOREM 2.4.7 $R_C(n) \le 1 - nq^k/(1+(n-2k+1)q^k + 2q(1-q^{k-1})/p)$.

Proof. Let A_i be $\overline{G}_i$ and Y be $\overline{G}$ in Lemma 2.4.6. Then

$$\Pr(\overline{G}_i|\overline{G}=j) = j/n.$$

Hence the condition of Lemma 2.4.6 is satisfied. From Corollary 2.4.5 and Lemma 2.4.6

$$R_C(1,n) \le 1 - E^2(\overline{G})/E(\overline{G}^2) = 1 - E(\overline{G})/E(\overline{G}|G_1) = 1 - nq^k/E(\overline{G}|G_1).$$

Let $I(\overline{G}_i)$ denote the indicator of $\overline{G}_i$. Then

$$E(\overline{G}|\overline{G}_1) = \sum_{i=1}^{n} E(I(\overline{G}_i)|\overline{G}_1)$$

$$\begin{aligned}
&= 1+\Big(\sum_{i=2}^{k} E(I(\overline{G}_i)|\overline{G}_1) + \sum_{i=n-k+2}^{n} E(I(\overline{G}_i)|\overline{G}_1)\Big) \\
&\quad + \sum_{i=k+1}^{n-k+1} E(I(\overline{G}_i)|\overline{G}_1) \\
&= 1+2\sum_{j=1}^{k-1} q^j + (n-2k+1)q^k \\
&= 1+2q(1-q^{k-1})/p + (n-2k+1)q^k.
\end{aligned}$$

The theorem then follows. ■

Let $E(d)$ denote the joint event that the line contains exactly d failed component and the line is failed. Chan, Chan and Lin [37] argued that when q_i are small, $\Pr(E(k)) + \Pr(E(k+1))$ is an adequate lower bound of $\overline{R}_L(1,n)$. They derived

$$\Pr(E(k)) = \left(\prod_{j=1}^{n} p_j\right) \sum_{i=1}^{n-k+1} \left(\prod_{j=i}^{i+k-1} \rho_j\right)$$

and

$$\Pr(E(k+1)) = \left(\prod_{j=1}^{n} p_j\right)\left(\sum_{i=1}^{n-k+1} \left(\prod_{j=i}^{i+k-1} \rho_j\right)\left(\sum_{\ell\notin[i,i+k-1]} \rho_\ell\right)\right). \quad (2.4.2)$$

The first term in (2.4.2) computes the probability of consecutive-k failed nodes plus one isolated failed node; the second computes that of consecutive-$(k+1)$ failed nodes.

Papastavridis [185] obtained following bounds from the generating functions.

THEOREM 2.4.8 *If $q < k/(k+1)$, then*

$$bm^{n+1} - e < R_L(n) < aM^{n+1} + e$$

and

$$m^n - (k-1)q^n < R_C(n) < M^n + (k-1)q^n,$$

where

$$m = 1 - \frac{pq^k}{(1-q^k)^k}, \quad M = 1 - pq^k, \quad a = \frac{m^k - q^k}{m^k - (k+1)pq^k},$$

$$b = \frac{M^k - q^k}{M^k - (k+1)pq^k}, \quad e = \frac{2(k-1)q^{n+2}}{p(k+(k+1)q)}.$$

Fu [78] gave the following result (with a slight improvement from Papastavridis and Koutras [198]).

THEOREM 2.4.9 *If $q_j < p_j$ for all j, then*

$$\prod_{i=k}^{n}\Big(1-\prod_{j=i-k+1}^{i} q_j\Big) \le R_L(1,n) \le \prod_{i=k}^{n}\Big(1-p_{i-k}\prod_{j=i-k+1}^{i} q_j\Big).$$

Proof. The LHS is just the IND version of Theorem 2.4.1. To prove the RHS, note that

$$R_L(1,n) = \prod_{i=k}^{n}\Pr(S_L(1,i)|S_L(1,i-1)) = \prod_{i=k}^{n}(1-\Pr(\overline{S}_L(1,i)|S_L(1,i-1))),$$

where

$$\begin{aligned}
&\Pr(\overline{S}_L(1,i)|S_L(1,i-1))\\
&\quad= \frac{\Pr(\overline{S}_L(1,i)\cap S_L(1,i-1))}{\Pr(S_L(1,i-1))}\\
&\quad= \frac{\Pr(\overline{S}_L(1,i)\cap S_L(1,i-1))}{\Pr(S_L(1,i-1)\cap s_{i-k})+\Pr(S_L(1,i-1)\cap \overline{s}_{i-k})}\\
&\quad\ge \frac{\Pr(\overline{S}_L(1,i)\cap S_L(1,i-1))}{\Pr(S_L(1,i-1)\cap s_{i-k})}\,\frac{1}{1+q_{i-k}/p_{i-k}}\\
&\quad= \frac{\Pr(S_L(1,i-k-1)\cap s_{i-k}\overline{s}_{i-k+1}\overline{s}_{i-k+2}\cdots\overline{s}_i)}{\Pr(S_L(1,i-1)\cap s_{i-k})}\,\frac{p_{i-k}}{1}\\
&\quad= p_{i-k}\prod_{j=i-k+1}^{i} q_j,
\end{aligned}$$

since

$$\begin{aligned}
\Pr(S_L(1,i-1)\cap \overline{s}_{i-k}) &= \Pr(S_L(1,i-1)|\overline{s}_{i-k})q_{i-k}\\
&\le \Pr(S_L(1,i-1)|s_{i-k})q_{i-k}\\
&= \Pr(S_L(1,i-1)\cap s_{i-k})q_{i-k}/p_{i-k}.
\end{aligned}$$

■

Fu [78] actually gave the weaker upper bound

$$R_L(1,n) \le \prod_{i=k}^{n}\left(1-(1-\rho_{i-k})\prod_{j=i-k+1}^{i} q_j\right).$$

Using Theorem 2.4.9, Fu obtained

THEOREM 2.4.10 *If*

$$\sum_{j=1}^{\infty}\frac{1}{j}\sum_{i=k}^{n}\Big(\prod_{j=i-k+1}^{i} q_j\Big)^j \to \underline{\theta} \text{ as } n \to \infty \quad \text{and}$$

$$\sum_{j=1}^{\infty}\frac{1}{j}\sum_{i=k}^{n}\Big(p_{i-k}\prod_{j=i-k+1}^{i} q_j\Big)^j \to \overline{\theta} \text{ as } n \to \infty,$$

then

$$e^{-\underline{\theta}} \le R(1,n) \le e^{-\overline{\theta}}.$$

Proof. Theorem 8.55 in Page 209 of Apostol [8] says that if $0 < u_i < 1$ for all i, then $\prod_{i=1}^{\infty}(1-u_i)$ and $\sum_{i=1}^{\infty} u_i$ both converge or diverge together. Now

$$\log\prod_{i=k}^{n}(1-u_i) = \sum_{i=k}^{n}\log(1-u_i) = -\sum_{j=1}^{\infty}\frac{1}{j}\sum_{i=k}^{n}(u_i)^j$$

holds for all n, and hence also holds as $n \to \infty$. Defining u_i as $\prod_{j=i-k+1}^{i} q_j$ or $p_{i-k}\prod_{j=i-k+1}^{i} q_j$ in Theorem 2.4.9 yields the desired result. ■

COROLLARY 2.4.11 *If* $\underline{\theta} = \overline{\theta} = \theta$, *then* $R(1,n)$ *is approximated by* $e^{-\theta}$.

Chapter 3

DESIGN OF OPTIMAL CONSECUTIVE SYSTEMS

For the design problem, we separate the notion of a component from its location in the system. Namely, we consider the moment that the n components are not in the system yet, and the problem is to assign them to the n locations, given that any component can go to any location, to maximize the system reliability.

3.1 OPTIMAL CONSECUTIVE-2 SYSTEMS

Derman, Lieberman and Ross [68] proposed two optimal design problems for linear consecutive-2 systems and solved one of them. Consider n components with reliabilities $p_1, p_2, \ldots, p_n$. Let $p_{[i]}$ denote the ith smallest p_j. In the *sequential assignment problem*, components are assigned one at a time to the system, and the state of component is learned as soon as it is connected to the system. In the *nonadaptive assignment problem*, the system is constructed all at once.

For the sequential assignment problem, Derman, Lieberman and Ross [68] proposed the following rule of assignment: Assign the least reliable component first. Afterwards, if the last assigned component works, assign the least reliable component in the remaining pool next; if the last assigned component is failed, assign the most reliable component in the remaining pool next. Call this rule DLR. Let L denote a line and let $p_x p_y L$ be obtained from L by adding two ordered components with reliabilities p_x and p_y before L.

LEMMA 3.1.1 *If* $p_i < p_j$ *and* $pR(L) > 0$, *then* $p_i R(p_j pL) < p_j R(p_i pL)$.

Proof. Note that

$$p_i R(p_j pL) = p_i\Big(p_j R(pL) + (1 - p_j) pR(L)\Big)$$

$$\begin{aligned} &= p_i p_j R(pL) + p_i pR(L) - p_i p_j pR(L) \\ &< p_i p_j R(pL) + p_j pR(L) - p_i p_j pR(L) \\ &= p_j\Big(p_i R(pL) + (1 - p_i)pR(L)\Big) \\ &= p_j R(p_i pL). \end{aligned}$$

■

THEOREM 3.1.2 *The assignment rule DLR is optimal.*

Proof. Suppose the last assigned component u is failed and an optimal algorithm A assigns p_i next while p_j, with $p_j > p_i$, is available. Let A' be the algorithm obtained from A by interchanging p_i and p_j in their assignments. Let v be the component assigned before p_j in A (v could be p_i). If v is failed, then A' succeeds if and only if A succeeds. Therefore, assume v works. Let A_1 denote the (could be empty) subsequence of A between p_i and p_j, and A_2 the subsequence after p_j. Then, by Lemma 3.1.1,

$$\begin{aligned} &\Pr(A \text{ works}|v \text{ works}) - \Pr(A' \text{ works}|v \text{ works}) \\ = \ &\Pr(A_1 \text{ works}|v \text{ works}) \Big(p_i R(p_j A_2) - p_j R(p_i A_2)\Big) < 0. \end{aligned}$$

The case u is working can be similarly proved. Finally, note that the n-line can be treated as an $(n+1)$-line starting with a pseudo component with reliability 1. So assigning the least reliable component first becomes a special application of the assignment rule following a working component. ■

For the nonadaptive assignment problem, Derman, Lieberman and Ross [68] conjectured the optimal line is

$$\tilde{L}_n = (p_{[1]}, p_{[n]}, p_{[3]}, p_{[n-2]}, \cdots, p_{[n-3]}, p_{[4]}, p_{[n-1]}, p_{[2]}),$$

where $p_{[i]}$ is the ith smallest element among $p_1, p_2, \ldots, p_n$. The conjecture was proved by Wei, Hwang and Sös [251] for two special cases that there are only two distinct values for p_i and there are three distinct values including 0 or 1. Hwang [101] extended this conjecture to the circular system as shown in Figure 3.1.

Note that any consecutive-k-out-of-n linear system can be formulated as a consecutive-k-out-of-$(n+1)$ circular system by setting $p_{n+1} = 1$. Therefore, the line conjecture holds if the cycle conjecture holds. Du and Hwang [70] proved the cycle conjecture, while Malon [172] claimed a proof of the line conjecture (however, see the end of this section for a critique). We will give a proof for the cycle which combines Malon's

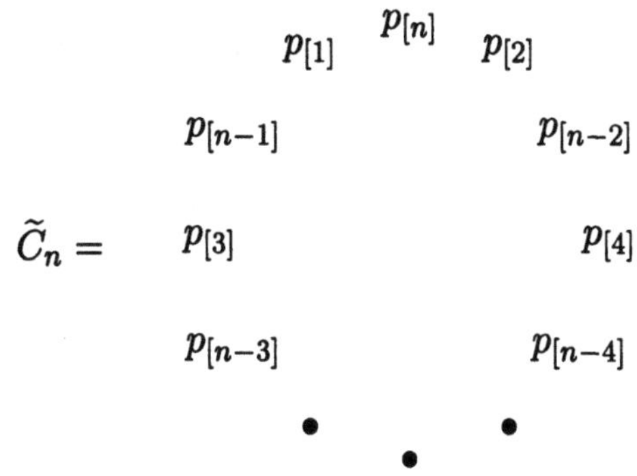

Figure 3.1.1. The optimal consecutive-2 circular system.

technique for the line and Du-Hwang's technique for the cycle, but is simpler than both.

Consider a sequence $(p_1, p_2, \ldots, p_\ell)$. For easier presentation, we assume that the p_i are distinct as the non-distinct case will follow by continuity (but uniqueness can be lost). For $i < j$, i is called a *singular index* if $p_i < p_j$ and $p_{i+1} < p_{j-1}$. A sequence is *singularity free* if it has no singular index.

LEMMA 3.1.3 *Let* $L = (p_1, p_2, \ldots, p_\ell)$ *denote a singularity free sequence. If* $p_1 < p_\ell$, *then* $\Pr(S(L)|s_1) < \Pr(S(L)|s_\ell)$.

Proof. The lemma is trivially true for $\ell = 1, 2$. The general ℓ case is proved by induction. Since L is singularity free, $p_2 \geq p_{\ell-1}$. Let $L' = (p_2, p_3, \ldots, p_{\ell-1})$. Then

$$\begin{aligned}
\Pr(S(L)|s_1\bar{s}_\ell) &= p_1 q_\ell \Pr(S(L')|s_{\ell-1}) \\
&\leq p_1 q_\ell \Pr(S(L')|s_2) \quad \text{(induction hypothesis)} \\
&< q_1 p_\ell \Pr(S(L')|s_2) \quad (\text{since } p_1 < p_\ell) \\
&= \Pr(S(L')|\bar{s}_1 s_\ell).
\end{aligned}$$

$$\begin{aligned}
\text{Therefore,}\quad \Pr(S(L)|s_1) &= \Pr(S(L)|s_1 s_\ell) + \Pr(S(L)|s_1\bar{s}_\ell) \\
&< \Pr(S(L)|s_1 s_\ell) + \Pr(S(L)|\bar{s}_1 s_\ell) \\
&= \Pr(S(L)|s_\ell).
\end{aligned}$$

■

A circular system is called *singularity free* if every subsequence is singularity free.

THEOREM 3.1.4 *An optimal consecutive-2 circular system is singularity free.*

Proof. Suppose to the contrary that an optimal consecutive-2 circular system C_n has a singular index $p_i < p_j$ and $p_{i+1} < p_{j-1}$. By shrinking the interval $(p_i, p_{i+1}, \ldots, p_{j-1}, p_j)$, if necessary, we may assume that $(p_{i+1}, p_{i+2}, \ldots, p_{j-2}, p_{j-1})$ is singularity free. Let G denote the subsequence $(p_{i+1}, p_{i+2}, \ldots, p_{j-2}, p_{j-1})$, and let H denote the subsequence $(p_{j+1}, p_{j+2}, \ldots, p_{i-2}, p_{i-1})$. Then C_n is as shown in Figure 3.1.1 (a).

H $\qquad$ H^{-1} $\qquad$ H

p_j $\quad$ p_i $\quad$ p_i $\quad$ p_j $\quad$ p_i $\quad$ p_j

G $\qquad$ G $\qquad$ G

(a) C_n $\qquad$ (b) C'_n $\qquad$ (c) C''_n

Figure 3.1.2. Three circular systems.

Let C'_n and C''_n be obtained from C_n as shown in Figure 3.1.1 (b) and (c), where $H^{-1} = (p_{i-1}, p_{i-2}, \ldots, p_{j+2}, p_{j+1})$. Clearly, if both i and j work or both fail, then C'_n and C''_n work if and only if C_n works. Therefore, it suffices to consider the case that one works and the other fail.

$$\begin{aligned}
\Pr(C'_n) - \Pr(C_n) &= q_i p_j \Pr(S(G)|s_{j-1}) \Pr(S(H)|s_{i-1}) + \\
&\quad p_i q_j \Pr(S(G)|s_{i+1}) \Pr(S(H)|s_{j+1}) - \\
&\quad q_i p_j \Pr(S(G)|s_{i+1}) \Pr(S(H)|s_{i-1}) - \\
&\quad p_i q_j \Pr(S(G)|s_{j-1}) \Pr(S(H)|s_{j+1}) \\
&= \Big(\Pr(S(G)|s_{j-1}) - \Pr(S(G)|s_{i+1})\Big) \\
&\quad \Big(q_i p_j \Pr(S(H)|s_{i-1}) - p_i q_j \Pr(S(H)|s_{j+1})\Big), \\
\Pr(C''_n) - \Pr(C_n) &= q_i p_j \Pr(S(G)|s_{j-1}) \Pr(S(H)|s_{j+1}) + \\
&\quad p_i q_j \Pr(S(G)|s_{i+1}) \Pr(S(H)|s_{i-1}) - \\
&\quad q_i p_j \Pr(S(G)|s_{i+1}) \Pr(S(H)|s_{i-1}) - \\
&\quad p_i q_j \Pr(S(G)|s_{j-1}) \Pr(S(H)|s_{j+1})
\end{aligned}$$

$$= (q_ip_j - p_iq_j)\Big(\Pr(S(G)|s_{j-1})\Pr(S(H)|s_{j+1}) - \Pr(S(G)|s_{i+1})\Pr(S(H)|s_{i-1})\Big).$$

Since

$$q_ip_j > p_iq_j \text{ and } \Pr(S(G)|s_{j-1}) > \Pr(S(G)|s_{i+1}),$$

it follows

$$\Pr(C'_n) > \Pr(C_n) \qquad \text{if } \Pr(S(H)|s_{i-1}) \geq \Pr(S(H)|s_{j+1}),$$
$$\Pr(C''_n) > \Pr(C_n) \qquad \text{if } \Pr(S(H)|s_{j+1}) \geq \Pr(S(H)|s_{i-1}).$$

Therefore, C_n is not optimal as assumed. ∎

COROLLARY 3.1.5 *$\widetilde{C}_n$ as shown in Figure* 3.1.1 *is the unique optimal consecutive-*2 *cycle.*

Proof. We show that $\widetilde{C}_n$ is the unique singularity free cycle. Note that node 1 must be adjacent to nodes n and $n-1$. Suppose node $x \in \{n, n-1\}$ is not adjacent to node 1. Then node 1 is adjacent to $y \notin \{n, n-1\}$. Then we have the singularity $x > y$ and $z > 1$, where z is a node adjacent to x. Similarly, node n must be adjacent to nodes 1 and 2, and node 2 must be adjacent to nodes n and $n-2$, and so on. ∎

COROLLARY 3.1.6 *$\widetilde{L}_n$ is the unique optimal consecutive-*2 *line.*

It should be noted that in Malon's proof of the linear system, inequalities similar to those given in the proof of Theorem 3.1.4 were given, but the inequalities are not strict. Then what was proved was that the optimality can be preserved by removing a particular singularity. However, since it was not shown that this removal reduces the total number of singularities, i.e., no new singularity is created, one couldn't conclude that the cycle with no singularity is optimal, let alone it is the optimal cycle. It is of interest to see whether there exists a procedure to obtain $\widetilde{C}_n$ from any cycle.

Note that the optimal dynamic consecutive-2 line, the optimal static consecutive-2 line and cycle all depend on the ranks of p_i, but not their actual values. Such an optimal system is called *invariant.* One would expect the remaining case, the optimal dynamic consecutive-2 cycle, is also invariant. Recently, Hwang and Pai [112] gave the surprising negative result. Let $G(i)$ denote the Derman, Lieberman and Ross algorithm given for the dynamic line except the first component assigned is node i. They showed that an invariant algorithm must be among $G(i)$: $\lceil n/2 \rceil \leq i \leq n-2$. They also showed that for $n = 8$, $G(6)$ is optimal if $p_i = p$, but $G(5)$ is optimal if $p_{[4]} > p_{[3]} = 0$.

3.2 INVARIANT CONSECUTIVE-k SYSTEMS

Malon [172] completely solved the invariant consecutive-k line problem.

THEOREM 3.2.1 *There exist invariant consecutive-k lines if and only if $k \in \{1, 2, n-2, n-1, n\}$. The invariant lines are given as below:*

(*any arrangement*)	*if* $k = 1$;
$(p_{[1]}, p_{[n]}, p_{[3]}, p_{n-2]}, \cdots, p_{[n-3]}, p_{[4]}, p_{n-1]}, p_{[2]})$	*if* $k = 2$;
$(p_{[1]}, p_{[4]},$ (*any arrangement*)$, p_{[3]}, p_{[2]})$	*if* $k = n-2$;
$(p_{[1]},$ (*any arrangement*)$, p_{[2]})$	*if* $k = n-1$;
(*any arrangement*)	*if* $k = n$.

Proof. The $k = 1, n$ case is trivial. The $k = 2$ case is proved in Corollary 3.1.6. We first prove that no invariant line exists for $n/2 \geq k > 2$.

Consider the case $p_{[1]} = \ldots = p_{[k-1]} = 0 < p_{[k]} = \ldots = p_{[n-1]} = 1-\epsilon < 1$, where $\epsilon \to 0$, and $p_{[n]} = 1$. Then the unique (up to equivalence by reversing) optimal line is placing $p_{[1]}, p_{[2]}, \cdots, p_{[k-1]}$ at the first $k-1$ positions and $p_{[n]}$ at position k, since this has unreliability $o(\epsilon^k)$ and any other case has unreliability $o(\epsilon^m)$ with $m < k$.

On the other hand, consider the case $p_{[1]} = \ldots = p_{[k-1]} = 0 < p_{[k]} = \ldots = p_{[n]} = 1-\epsilon < 1$ where $\epsilon \to 0$. Then, placing $p_{[1]}, \cdots, p_{[k-1]}$ in consecutive positions is worse than having them separated by exactly two other p_i's, since the former has unreliability $o(\epsilon)$ and the later has unreliability $o(\epsilon^2)$. This contradicts the conclusion that $p_{[1]}, p_{[2]}, \cdots, p_{[k-1]}$ must be placed at the first $k-1$ positions.

Next consider $n/2 < k < n$. Note that every cutset contains components $n-k+1, n-k+2, \ldots, k-1, k$. Therefore $L(1,n)$ is failed if and only if these components are all failed; and among the remaining components, some $n-k$ consecutive components are failed (nodes $n-k$ and $k+1$ are considered consecutive). Clearly, an invariant line must have $\{p_{n-k+1}, p_{n-k+2}, \cdots, p_k\} = \{p_{[n]}, p_{[n-1]}, \cdots, p_{[2n-2k+1]}\}$, and an invariant consecutive-$(n-k)$-out-of-$(2n-2k)$ line with reliabilities $\{p_{[1]}, p_{[2]}, \cdots, p_{[2n-2k]}\}$. But such invariant lines exist only for $n-k = 1$ or 2 (as given in the theorem). ■

Hwang [105] extended Malon's result to the cycle.

THEOREM 3.2.2 *There exist invariant consecutive-k cycles if and only if $k \in \{1, 2, n-2, n-1, n\}$. For $k \in \{1, n-1, n\}$, any cycle is optimal. For $k \in \{2, n-2\}$, the unique invariant cycle is $\widetilde{C}_n$.*

Proof. Only the $k = n-2$ case needs to be discussed. A consecutive-$(n-2)$ cycle is failed if and only if at most one node works or exactly two nodes work, but they are adjacent. The sequencing can only affect the latter case. Therefore, we want to minimize

$$\sum_{i=1}^{n} p_i p_{i+1} q_{i+2} \cdots q_{i-1} = \prod_{j=1}^{n} q_j \sum_{i=1}^{n} \lambda_i \lambda_{i+1},$$

or simply, to minimize $\sum_{i=1}^{n} \lambda_i \lambda_{i+1}$, where $\lambda_i = p_i/q_i$.

Suppose the cycle $C = (\lambda_1, \lambda_2, \ldots, \lambda_n)$ is not singularity free, say, $\lambda_i > \lambda_j$ and $\lambda_{i+1} > \lambda_{j-1}$. Define $f(C) = \sum_{i=1}^{n} \lambda_i \lambda_{i+1}$. Let C' be the cycle obtained from C by reversing the subsequence $(\lambda_{i+1}, \lambda_{i+2}, \ldots, \lambda_{j-1})$,

$$\begin{aligned} f(C) - f(C') &= \lambda_i \lambda_{i+1} + \lambda_{j-1}\lambda_j - \lambda_i \lambda_{j-1} - \lambda_j \lambda_{i+1} \\ &= (\lambda_i - \lambda_j)(\lambda_{i+1} - \lambda_{j-1}) \\ &> 0. \end{aligned}$$

It follows that an optimal cycle is singularity free, and $\widetilde{C}_n$ is the unique one. Not that λ_i is increasing in p_i. Hence $\widetilde{C}_n$ for $\{\lambda_i\}$ is the same $\widetilde{C}_n$ for $\{p_i\}$.

Next we show that no invariant cycle exists for $3 \le k \le n-3$. Consider the case $p_{[n]} = 1$. Then an invariant consecutive-k-out-of-n cycle implies an invariant consecutive-k-out-of-$(n-1)$ line, which by Theorem 3.2.1 does not exist except perhaps for $k = n-3$. But then the invariant line must be in the form $(p_{[1]}, p_{[4]}, (\text{any permutation of } p_{[5]} \text{ to } p_{[n-1]}), p_{[3]}, p_{[2]})$, which implies the invariant cycle has $p_{[n]}$ between $p_{[1]}$ and $p_{[2]}$. A few examples were then given by Hwang [105] to show the nonoptimality for such a sequencing. ∎

Sometimes there is no universal invariant system, but there are some local invariant assignments. Such knowledge reduces the number of system we need to search as candidates of optimal systems.

Tong [245] gave a sufficient condition for interchanging two adjacent components.

THEOREM 3.2.3 *Assume that* $(n-1)/2 \le k \le n-2$ *and* $p_1 > p_2$. *Let* C' *be obtained from* $C = (p_1, p_2, \ldots, p_n)$ *by interchanging* p_1 *and* p_2. *Then* $R(C') \ge R(C)$ *if and only if*

$$\left(\prod_{i=3}^{k+1} q_i\right) p_{k+2} - p_{n-k+1}\left(\prod_{i=n-k+2}^{n} q_i\right) \le 0.$$

Proof. It is easily seen that

$$R(C') - R(C) = (p_1 q_2 - p_2 q_1)\left(\left(\prod_{i=3}^{k+1} q_i\right) p_{k+2} - p_{n-k+1}\left(\prod_{i=n-k+2}^{n} q_i\right)\right)$$

$$= (p_1 - p_2)\left(\left(\prod_{i=3}^{k+1} q_i\right)p_{k+2} - p_{n-k+1}\left(\prod_{i=n-k+2}^{n} q_i\right)\right).$$

The theorem follows immediately. ■

Component i is said to be *criticality more important* than component j if $\{j\} \cup S \in C$ implies $\{i\} \cup S \in C$ for every set S which contains neither i nor j. They are *criticality equivalent* if each is more critical than the other. Clearly, a criticality more important component should always be assigned a larger $p_{[i]}$.

THEOREM 3.2.4 *Node $i+1$ is criticality more important than node i for $1 \le i \le k-1$. For $n < 2k$, nodes i and j are criticality equivalent for $n-k+1 \le i < j \le k$.*

COROLLARY 3.2.5 *In an optimal consecutive-k line, $p_1 \le p_2 \le \ldots \le p_k$.*

Corollary 3.2.5 was independently given by Tong [244] and Malon [172] for $n < 2k$. Kuo, Zhang and Zuo [150] observed that it holds in general.

3.3 THE BIRNBAUM IMPORTANCE

The *Birnbaum (reliability) importance* of component i in system $x \in \{L, C\}$ is defined as

$$I_x(i) = \frac{\partial R_x(p_1, p_2, \ldots, p_n)}{\partial p_i}.$$

If necessary, we use $I_x(i, n)$ to denote $I_x(i)$ with n fixed. It is the rate at which system reliability improves when the reliability of component i improves. Griffith and Govindarajulu [96], and Papastavridis [188] independently, first studied Birnbaum importance for the consecutive line.

THEOREM 3.3.1 $I_L(i) = \Big(R_L(1, i-1)R_L(i+1, n) - R_L(1, n)\Big)/q_i$ *and* $I_C(i) = \Big(R_L(p_{i+1}, p_{i+2}, \ldots, p_{i-1}) - R_C(1, n)\Big)/q_i$.

Proof. Since,

$$\begin{aligned} R_x(1,n) &= p_i R_x(p_1, \ldots, p_{i-1}, 1, p_{i+1}, \ldots, p_n) + \\ &\quad (1-p_i) R_x(p_1, \ldots, p_{i-1}, 0, p_{i+1}, \ldots, p_n), \end{aligned}$$

we have

$$\frac{\partial R_x(1,n)}{\partial p_i} = R_x(p_1, \ldots, p_{i-1}, 1, p_{i+1}, \ldots, p_n) -$$

$$\begin{aligned}
& R_x(p_1,\ldots,p_{i-1},0,p_{i+1},\ldots,p_n) \\
= \;& R_x(p_1,\ldots,p_{i-1},1,p_{i+1},\ldots,p_n) - \\
& \Big(R_x(1,n) - p_i R_x(p_1,\ldots,p_{i-1},1,p_{i+1},\ldots,p_n)\Big)/q_i \\
= \;& \Big(R_x(p_1,\ldots,p_{i-1},1,p_{i+1},\ldots,p_n) - R_x(1,n)\Big)/q_i.
\end{aligned}$$

The formula for $I_L(i)$ then follows from

$$R_L(p_1,\ldots,p_{i-1},1,p_{i+1},\ldots,p_n) = R_L(1,i-1)R_L(i+1,n)$$

and the formula for $I_C(i)$ from

$$R_C(p_1,\ldots,p_{i-1},1,p_{i+1},\ldots,p_n) = R_L(p_{i+1},p_{i+2},\ldots,p_{i-1}).$$

■

COROLLARY 3.3.2 *For the IID model,* $I_L(i) = \Big(R_L(i-1)R_L(n-i) - R_L(n)\Big)/q$ *and* $I_C(i) = \Big(R_L(n-1) - R_C(n)\Big)/q$.

The Birnbaum importance function observes the same recursive equations as observed by the reliability function. Thus we have

THEOREM 3.3.3 *For IID model,*
(i) $I_L(i,n) = I_L(i,n-1) - pq^k I_L(i,n-k-1)$ *if* $n-i \geq k+1$,
(ii) $I_L(i,n) = I_L(i-1,n-1) - pq^k I_L(i-k-1,n-k-1)$ *if* $i-1 \geq k+1$.

Proof.

$$\begin{aligned}
\text{(i) } qI_L(i,n) &= R_L(i-1)R_L(n-i) - R_L(n) \\
&= R_L(i-1)(R_L(n-i-1) - pq^k R_L(n-i-k-1)) - \\
&\quad (R_L(n-1) - pq^k R_L(n-k-1)) \\
&= (R_L(i-1)R_L(n-i-1) - R_L(n-1)) - \\
&\quad pq^k(R_L(i-1)R_L(n-i-k-1) - R_L(n-k-1)) \\
&= qI_L(i,n-1) - pq^k qI_L(i,n-k-1). \\
\text{(ii) } qI_L(i,n) &= RL(i-1)R_L(n-i) - R_L(n) \\
&= (R_L(i-2) - pq^k R_L(i-k-2))R_L(n-i) - \\
&\quad (R_L(n-1) - pq^k R_L(n-k-1)) \\
&= (R_L(i-2)R_L(n-i) - R_L(n-1)) - \\
&\quad pq^k(R_L(i-k-2)R_L(n-i) - R_L(n-k-1)) \\
&= qI_L(i-1,n-1) - pq^k qI_L(i-k-1,n-k-1).
\end{aligned}$$

■

The comparison of Birnbaum importance under the IND model is valid only for the underlying set $(p_1, p_2, \ldots, p_n)$. On the other hand, one can hardly expect that $I_L(i) \geq I_L(j)$ for all $(p_1, p_2, \ldots, p_n)$, except when i is criticality more important than j. Therefore the IID model is the suitable one if the focus is on comparing the positions by neutralizing the differences in p_i. Even for the IID model, Hwang, Cui, Chang and Lin [110] showed that the comparison is not independent of p.

To compare importance for general k, Zuo [266] gave the following lemma for the IND version.

LEMMA 3.3.4 *If $n \geq k \geq 2$ and $1 > p_i > 0$ for all i, then for any i we have*

$$R_L(1, i-1)p_i R_L(i+1, n) < R_L(1, n) < R_L(1, i)R_L(i+1, n).$$

Proof. The second inequality is true since the system works only if all disjoint parts work. The first inequality is true since the left hand side represents a case that the system works. ■

COROLLARY 3.3.5 *If $n \geq k \geq 2$ and $1 > p > 0$, then*

$$pR_L(i-1)R_L(n-i) < R_L(n) < R_L(i)R_L(n-i).$$

By symmetry, $I_x(i)$ for a circular system or a line system with $k = 1$ or n must be a constant under the IID model. Zuo and Kuo [270] studied the $k = 2$ case. By symmetry, only $I_L(i)$ for $i \leq n/2$ need be given.

LEMMA 3.3.6 *For $k = 2$ and $2 \leq m \leq n-2$,*

$$R_L(n) = R_L(m)R_L(n-m) - p^2q^2R_L(m-2)R_L(n-m-2).$$

Proof. Partition an n-line into an m-line and an $(n-m)$-line in that order. Then the only case the latter two lines both work but not the former line is when the last component of the m-line and the first component of the $(n-m)$-line both fail. But for the m-line and the $(n-m)$-line to work, we must also have the second last component of the m-line working and the remaining $(m-2)$-line working, as well as the second component of the $(n-m)$-line working and the remaining $(n-m-2)$-line working. ■

We now define

$$J(i) = R_L(i-1)R_L(n-i).$$

If necessary, we use $J(i, n)$ to denote $J(i)$ with n fixed. Clearly, comparing $I_L(i)$ (or $I_L(i, n)$) is equivalent to comparing $J(i)$ (or $J(i, n)$).

THEOREM 3.3.7 *Consider the consecutive-2 line under the IID model. Then*

$$\begin{array}{ll} I_L(2i) > I_L(2i-1) & \text{for } 2i \le (n+1)/2, \\ I_L(2i) < I_L(2i-2) & \text{for } 2i \le (n+1)/2, \\ I_L(2i+1) > I_L(2i-1) & \text{for } 2i+1 \le (n+1)/2. \end{array}$$

Proof. Theorem 3.3.7 is easily verified for $n \le 4$. We prove the general n case by induction. Using Corollary 2.1.2 and Lemma 3.3.6,

$$\begin{aligned} J(1) &= R_L(0)R_L(n-1) = R_L(n-1), \\ J(2) &= R_L(1)R_L(n-2) = R_L(n-1) + pq^2R_L(n-4), \\ J(3) &= R_L(2)R_L(n-3) = R_L(n-1) + p^2q^2R_L(n-5), \\ J(4) &= R_L(3)R_L(n-4) = R_L(n-1) + p^2q^2R_L(n-6). \end{aligned}$$

By noting that $R_L(m)$ is decreasing in m, we obtain $J(4) > J(3)$. By setting $i = 1$ and $n = n-4$ in the first inequality of Corollary 3.3.5, we have $R_L(n-4) > pR_L(n-5)$. Thus

$$J(2) > J(4) > J(3) > J(1).$$

Consider the general case. Assume $m + 3 \le (n+1)/2$.

$$\begin{aligned} J(m) &= R_L(m-1)R_L(n-m), \\ J(m+1) &= R_L(m)R_L(n-m-1), \\ J(m+2) &= R_L(m+1)R_L(n-m-2), \\ &= R_L(n-1) + p^2q^2R_L(m-1)R_L(n-m-4), \\ J(m+3) &= R_L(n-1) + p^2q^2R_L(m)R_L(n-m-5), \\ J(m+4) &= R_L(n-1) + p^2q^2R_L(m+1)R_L(n-m-6). \end{aligned}$$

Thus, if

$$J(m) \gtreqless J(m+1), \text{i.e., } R_L(m-1)R_L(n-m) \gtreqless R_L(m)R_L(n-m-1),$$

then by substituting n with $n-4$ we have

$$R_L(m-1)R_L(n-m-4) \gtreqless R_L(m)R_L(n-m-5), \text{i.e., } J(m+2) \gtreqless J(m+3).$$

Also, if

$$J(m) \gtreqless J(m+2), \text{i.e., } R_L(m-1)R_L(n-m) \gtreqless R_L(m+1)R_L(n-m-2),$$

then by substituting n with $n-4$ we have

$$R_L(m-1)R_L(n-m-4) \gtreqless R_L(m+1)R_L(n-m-6), \text{i.e., } J(m+2) \gtreqless J(m+4).$$

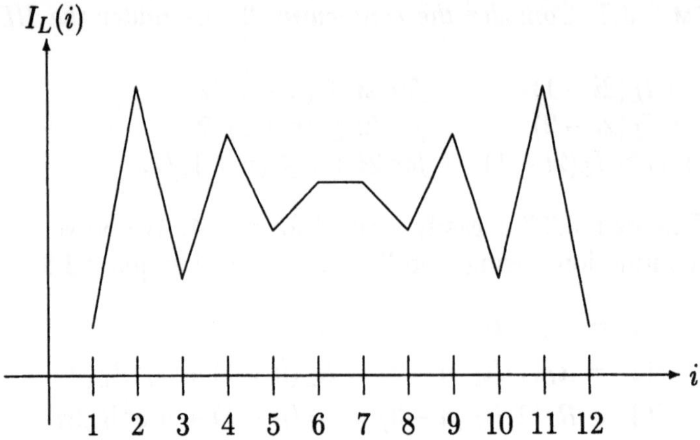

Figure 3.3.1. $I_L(i)$ for the consecutive-2-out-of-12 line.

The theorem then follows from induction. ■

Not much is known for general k. Clearly, criticality more important implies Birnbaum more important. Thus from Theorem 3.2.4, we have

THEOREM 3.3.8 *For the consecutive-k line under IID model,*

$$I_L(1) < I_L(2) < \ldots < I_L(k) \text{ if } n \geq 2k;$$

$$I_L(n-k+1) = I_L(n-k+2) = \ldots = I_L(k) \text{ if } n < 2k.$$

Theorem 3.3.8 was first observed by Kuo, Zhang and Zuo [150] with a different argument.

THEOREM 3.3.9 *Consider a consecutive-k line under the IID model. Then* $I_L(1) \leq I_L(i)$ *for all* $i \leq n/2$ *and* $I_L(k) > I_L(k+1)$ *for* $n > 2k$.

Proof. For $i \leq n/2$,

$$J(1) = R_L(n-1) < R_L(i-1)R_L(n-i) = J(i).$$

For $n > 2k$,

$$\begin{aligned}
J(k+1) &= R_L(k)R_L(n-k-1) \\
&= R_L(k)\Big(R_L(n-k) + pq^k R_L(n-2k-1)\Big) \\
&= (1-q^k)R_L(n-k) + pq^k R_L(k)R_L(n-2k-1) \\
&= R_L(n-k) - q^k\Big(R_L(n-k) - pR_L(k)R_L(n-2k-1)\Big) \\
&< R_L(n-k) \quad \text{by Corollary 3.3.5} \\
&= J(k).
\end{aligned}$$

■

Zuo [266] proved the first part of Theorem 3.3.9 for $I_L(1) \leq I_L(k+1)$ as well as the second part, which generalized a result of Zakaria, David and Kuo [262]. Zuo also proved $I_L(k) \geq I_L(2k)$.

Chang, Cui and Hwang [39] (also see Chang, Hwang and Cui [41] for some disclaims) gave a method to compare $I_L(i)$ with $I_L(i+1)$.

LEMMA 3.3.10 *$I_L(i,n) \gtreqless I_L(i+1,n)$ if and only if $I_L(i-k,n-k) \gtreqless I_L(i+1,n-k)$.*

Proof. By Corollary 2.1.13,

$$R_L(n) = R_L(n-1) - pq^k R_L(n-k-1).$$

Hence

$$R_L(n-1)/R_L(n) = 1 + pq^k R_L(n-k-1)/R_L(n).$$

Let the symbol $\propto$ denote that both sides have the same sign. Then for $k+1 \leq i+1 \leq (n+1)/2$,

$$\begin{aligned} & I_L(i,n) - I_L(i+1,n) \\ \propto\ & R_L(i-1)R_L(n-i) - R_L(i)R_L(n-i-1) \\ \propto\ & R_L(i-1)/R_L(i) - R_L(n-i-1)/R_L(n-i) \\ \propto\ & R_L(i-k-1)/R_L(i) - R_L(n-i-k-1)/R_L(n-i) \\ \propto\ & R_L(i-k-1)R_L(n-i) - R_L(i)R_L(n-i-k-1) \\ \propto\ & I_L(i-k,n-k) - I_L(i+1,n-k). \end{aligned}$$

■

Thus the comparison of $I_L(i,n)$ and $I_L(i+1,n)$ is reduced to a comparison of $I_L(i-k,n-k)$ with $I_L(i+1,n-k)$. If $i+1 > (n-k+1)/2$, subtract $i' = n-k-i$ for $i+1$. If $i-k \geq k$, the same reduction can repeat. Eventually, if node i is transformed to node 1, or node $i+1$ to node k, then we use the fact $I_L(1) \leq I_L(i)$ and $I_L(k) \geq I_L(i)$ (to be proved in Corollary 3.3.15) for all i to obtain

THEOREM 3.3.11 *$I_L(2k+1) < I_L(2k)$, $I_L(k+1) < I_L(k+2)$ and $I_L(2k-1,4k-1) < I_L(2k,4k-1)$.*

Recently, Chang, Chen and Hwang [43] extended the results in Theorem 3.3.11 to

THEOREM 3.3.12 *(i) $I_L((t-2)k-1,tk-1) < I_L((t-2)k,tk-1)$ for $t \geq 3$. (ii) $I_L(3k+1,6k+1) < I_L(3k,6k+1)$.*

Proof. (i) By Lemma 3.3.10,

$$I_L((t-2)k-1, tk-1) < I_L((t-2)k, tk-1)$$

if and only if

$$I_L((t-3)k-1, (t-1)k-1) < I_L((t-2)k, (t-1)k-1) = I_L(k, (t-1)k-1),$$

where the last inequality is due to the symmetry between k and $(t-2)k$ when $n = (t-1)k-1$. (i) now follows from Corollary 3.3.15.

(ii) By Lemma 3.3.10, (ii) is true if

$$I_L(2k, 5k+1) > I_L(3k+1, 5k+1) = I(2k+1, 5k+1),$$

which was proved in Theorem 3.3.11. ∎

Chang, Chen and Hwang also showed that

$$I_L(tk+1, 2tk+1) < I_L(tk, 2tk+1)$$

does not hold for $t = 4$.

Zuo and Kuo [270] proposed to use Birnbaum importance to construct a consecutive system when p_i are known. Given an initial sequence, compute $I_L(i)$, starting with the least reliable component, compare its importance with the importance of other components in order of reliability. If a violation is discovered, interchange the two and recompute the system reliability. Preserve the swapping if and only if the system reliability is improved. Continue until no more swapping is possible. Several initial sequences were offered as well as two different ways to proceed. Note that the procedure must end since the system reliability increases at each iteration.

Malinowski and Preuss [168] proposed the heuristic of interchanging two components over $\binom{n}{2}$ possible pairs and select the interchange which maximize the system reliability. If it is an improvement over the system before change, make the change and continue the same, otherwise the system stops. Note that each step requires $O(n^3)$ time, and it is not known how many steps are needed.

Recently, Hwang [109] defined a new importance measure. Component i is said to be H *more important* than component j, written $H(i) \geq H(j)$, if for every $d = k, k+1, \ldots, n$, $|\mathrm{CS}_{i,d}|$, the number of d-cutsets containing i is never fewer than $|\mathrm{CS}_{j,d}|$. Clearly, criticality more important implies H more important. On the other hand, Hwang proved

LEMMA 3.3.13 *$H(i) \geq H(j)$ implies $I_L(i) \geq I_L(j)$ under the IID model for all p.*

Proof. A *pathset* is a set of nodes such that the system works if all of them work. Let $PS_{i,d}$ and $PS_{\bar{i},d}$ denote the set of pathsets containing and not containing i, respectively.

$$\begin{aligned} I_L(i) &= R_L(p^{i-1}, 1, p^{n-i}) - R_L(p^{i-1}, 0, p^{n-i}) \\ &= \sum_{d \geq k} (|PS_{i,d}| - |PS_{\bar{i},d}|) p^{n-d} q^d \\ &= \sum_{d \geq k} (2|PS_{i,d}| - |PS_d|) p^{n-d} q^d, \end{aligned}$$

where PS_d is the set of d-pathsets. Note that S is a cutset if and only if the component of S is not a pathset. Furthermore, the complementary set does not contain i and has size $n - d$. Thus

$$|CS_{i,d}| + |PS_{\bar{i},n-d}| = 2^{n-1}.$$

It follows for each d

$$\begin{aligned} |CS_{i,d}| \geq |CS_{j,d}| &\Rightarrow |PS_{\bar{i},n-d} \leq |PS_{\bar{j},n-d}| \\ &\Rightarrow |PS_{i,n-d} \geq |PS_{j,n-d}|. \end{aligned}$$

Hence $I_L(i) \geq I_L(j)$. ■

Note that one cannot use computation to verify $I_L(i) \geq I_L(j)$ for all p since there is an infinite number of them. But for any finite system, we can verify $H(i) \geq H(j)$ since d is bounded by n. And Lemma 3.3.13 shows the surprising consequence that if $H(i) \geq H(j)$ is verified, then the previously impossible-to-verify relation $I_L(i) \geq I_L(j)$ is also verified.

Chang, Cui and Hwang [39] proved

THEOREM 3.3.14 *$H(k) > H(i)$ for all $i \leq (n+1)/2$.*

Proof. Let $f_i(d, n)$ denote $|CS_{i,d}|$ for given n. By Theorem 3.2.4,

$$f_k(d, n) \geq f_{k-1}(d, n) \geq \ldots \geq f_1(d, n).$$

Since the inequalities are strict for $d = k$, Theorem 3.3.14 holds for $n = 2k - 1$ and $2k$. We prove the general case by induction on n.

$$f_k(d, n) = \sum_{j=0}^{k-1} f_k(d - j, n - j - 1) + \binom{n-k-1}{d-k-1}.$$

The jth term in the sum corresponds to the case that there are exactly j consecutive failed components at the end; the last term to the case that the last k components at the end and component k are all failed. Similarly, for $i \leq (n+1)/2$,

$$f_i(d,n) = \sum_{j=0}^{k-1} f_i(d-j, n-j-1) + \binom{n-k-1}{d-k-1}.$$

Note that if $i > (n-j)/2$, then we can substitute i with $i' = n-j-i$ by symmetry. Furthermore, if k' substitutes for k in $f_k(d-j, n-j-1)$, then $n-j-1 < 2k-1$. Hence

$$\begin{aligned} f_k(d-j,n-j-1) &= f_{n-j-k}(d-j,n-j-1) \\ &> f_{n-j-i}(d-j,n-j-1) \\ &= f_i(d-j,n-j-1) \quad \text{for } k < i. \end{aligned}$$

Therefore,

$$\begin{aligned} f_k(d,n) &= \sum_{j=0}^{k-1} f_k(d-j,n-j-1) + \binom{n-k-1}{d-k-1} \\ &> \sum_{j=0}^{k-1} f_i(d-j,n-j-1) + \binom{n-k-1}{d-k-1} = f_i(d,n), \end{aligned}$$

since by induction

$$f_k(d-j,n-j-1) \geq f_i(d-j,n-j-1)$$

and the inequality is strict for some j. ■

COROLLARY 3.3.15 *If $i \neq k$, then $I_L(k) > I_L(i)$.*

3.4 HALF-LINE IMPORTANCE

Since it is difficult to compare Birnbaum importance, not even under the IID model, Chang, Chen and Hwang [43] proposed the *half-line importance* I^h which required $I_L(i) > I_L(j)$ only for all $p \geq 1/2$. They justified this condition by noting that it holds in most practical cases. All results in this section are taken from that paper.

First, a result of Lin, Kuo and Hwang [157] is extended from $p = 1/2$ to general p.

LEMMA 3.4.1 *Suppose $i \geq k+1$. Then* (i) *for $0 < j-i < k$,*

$$J(j,n) - J(i,n) = pq^k \sum_{s=k+1-j+i}^{k} \big(J(i, n-j+i-s) - J(i-s, n-j+i-s)\big);$$

(ii) *for* $j-i \geq k$ *and* $n-j \geq k$, $J(j,n) - J(i,n) =$

$$p^2q^k \sum_{s=2}^{k} R(j-i-s) \sum_{r=k+2-s}^{k} q^{r+s-k-2}\Big(J(i,n-j-r+i) - J(i-r,n-j-r+i)\Big).$$

Proof. (i) By Theorem 2.1.5,

$$\begin{aligned}
&J(j,n)\\
&= R_L(j-1)R_L(n-j)\\
&= \Big(R_L(i-1)R_L(j-i) - pq^k \sum_{s=k+1-j+i}^{k} R_L(i-1-s)\Big)R_L(n-j)\\
&= R_L(i-1)R_L(j-i)R_L(n-j) - pq^k \sum_{s=k+1-j+i}^{k} R_L(i-1-s)R_L(n-j)\\
&= R_L(i-1)\Big(R_L(n-i) + pq^k \sum_{s=k+1-j+i}^{k} R_L(n-j-s)\Big) -\\
&\quad pq^k \sum_{s=k+1-j+i}^{k} R_L(i-1-s)R_L(n-j)\\
&= J(i,n) + pq^k \sum_{s=k+1-j+i}^{k} \Big(J(i,n-j-s+i) - J(i-s,n-j-s+i)\Big).
\end{aligned}$$

(ii) Also,

$$\begin{aligned}
&J(j,n)\\
&= R_L(j-1)R_L(n-j)\\
&= \Bigg(R_L(i-1)R_L(j-i) -\\
&\quad p^2q^k \sum_{s=2}^{k} \Big(R_L(j-i-s) \sum_{r=k+2-s}^{k} q^{r+s-k-2} R_L(i-1-r)\Big)\Bigg) R_L(n-j)\\
&= R_L(i-1)R_L(j-i)R_L(n-j) -\\
&\quad p^2q^k \sum_{s=2}^{k} \Big(R_L(j-i-s) \sum_{r=k+2-s}^{k} q^{r+s-k-2} R_L(i-1-r)R_L(n-j)\Big)\\
&= R_L(i-1)\Bigg(R_L(n-i) +\\
&\quad p^2q^k \sum_{s=2}^{k} \Big(R_L(j-i-s) \sum_{r=k+2-s}^{k} q^{r+s-k-2} R_L(n-j-r)\Big)\Bigg) -
\end{aligned}$$

$$p^2q^k\sum_{s=2}^{k}\Big(R_L(j-i-s)\sum_{r=k+2-s}^{k}q^{r+s-k-2}R_L(i-1-r)R_L(n-j)\Big)$$

$$= J(i,n)+p^2q^k\sum_{s=2}^{k}\Bigg(R_L(j-i-s)$$

$$\sum_{r=k+2-s}^{k}q^{r+s-k-2}\Big(J(i,n-j-r+i)-J(i-r,n-j-r+i)\Big)\Bigg).$$

■

By using Corollary 2.1.4, another result of Lin, Kuo and Hwang [157] is extended from $p = 1/2$ to general p.

LEMMA 3.4.2 (i) $I_L(i,n)=\sum_{j=1}^{k}pq^{j-1}I_L(i-j,n-j)$ *for* $i \geq k+1$.

(ii) $I_L(i,n)=\sum_{j=1}^{k}pq^{j-1}I_L(i,n-j)$.

Proof. (i) For $j = 1, 2, \ldots, k$, we have

$$pq^{j-1}I_L(i-j,n-j)=pq^{j-1}\Big(R_L(i-k-1)R_L(n-i)-R_L(n-k)\Big)/q.$$

Summing up,

$$\sum_{j=1}^{k}pq^{j-1}I_L(i-j,n-j)$$

$$= \Big(R_L(n-i)\sum_{j=1}^{k}pq^{j-1}R_L(i-1-j)-\sum_{j=1}^{k}pq^{j-1}R_L(n-j)\Big)/q$$

$$= \Big(R_L(n-i)R_L(i-1)-R_L(n)\Big)/q$$

$$= I_L(i,n).$$

(ii) For $j = 1, 2, \ldots, k$, we have

$$pq^{j-1}I_L(i,n-j)=pq^{j-1}\Big(R_L(i-1)R_L(n-i-j)-R_L(n-j)\Big)/q.$$

Summing up,

$$\sum_{j=1}^{k}pq^{j-1}I_L(i,n-j)$$

$$= \Big(R_L(i-1)\sum_{j=1}^{k}pq^{j-1}R_L(n-i-j)-\sum_{j=1}^{k}pq^{j}R_L(n-j)\Big)/q$$

$$= \Big(R_L(i-1)R_L(n-i) - R_L(n)\Big)/q$$
$$= I_L(i,n).$$

■

COROLLARY 3.4.3 *If* $s \geq 0$, *then* $I_L(i+s,n) - I_L(i,n)$

$$= \sum_{j=1}^{k} pq^{j-1}\Big(I_L(i+s-j,n-j) - I_L(i,n-j)\Big) \textit{ for } i+s \geq k+1,$$
$$= \sum_{j=1}^{k} pq^{j-1}\Big(I_L(i+s,n-j) - I_L(i-j,n-j)\Big) \textit{ for } i \geq k+1,$$
$$= \sum_{j=1}^{k} pq^{j-1}\Big(I_L(i+s-j,n-j) - I_L(i-j,n-j)\Big) \textit{ for } i \geq k+1,$$
$$= \sum_{j=1}^{k} pq^{j-1}\Big(I_L(i+s,n-j) - I_L(i,n-j)\Big).$$

THEOREM 3.4.4 $I^h(k-1) < I^h(k+1)$.

Proof. $J^h(k+1) - J^h(k-1)$

$$\begin{aligned}
&= (1-q^k)R_L(n-k-1) - \\
&\quad \Big(R_L(n-k-1) - pq^kR_L(n-2k) - pq^kR_L(n-2k-1)\Big) \\
&= q^k\Big(pR_L(n-2k) + pR_L(n-2k-1) - R_L(n-k-1)\Big) \\
&\geq q^k\Big(R_L(n-2k)/2 + R_L(n-2k-1)/2 - R_L(n-k-1)\Big) \\
&\qquad \text{(by the fact that } p \geq 1/2) \\
&> 0 \ \text{(by the fact that } R_L(u) > R_L(v) \text{ for } u < v).
\end{aligned}$$

■

COROLLARY 3.4.5 $I^h(1) < I^h(2) < \ldots < I^h(k-1) < I^h(k+1)$.

Proof. The last inequality is from Theorem 3.4.4. The other inequalities are from Theorem 3.3.8. ■

THEOREM 3.4.6 $I^h(k+1) < I^h(i)$ *for all* $k+1 < i$.

Proof. The case $k+1 < i \leq 2k$ follows from Lemma 3.4.1 (i) and Corollary 3.4.5. The case $2k < i$ follows from Lemma 3.4.1 (ii) and Corollary 3.4.5. ■

COROLLARY 3.4.7 $I^h(2k+2) > I^h(2k+1)$.

Proof. By Lemma 3.3.10,

$$I^h(2k+2,n) > I^h(2k+1,n) \text{ if } I^h(2k+2,n-k) > I^h(k+1,n-k).$$

Corollary 3.4.7 follows immediately from Theorem 3.4.6. ■

THEOREM 3.4.8 $I^h(i) < I^h(i+1)$ *for* $k+1 \leq i \leq 2k-1$.

Proof. By Theorem 3.3.11, Theorem 3.4.8 holds for $i = k+1$. We prove the general case by induction. By Corollary 3.4.5 and Theorem 3.4.6,

$$I^h(i+1) > I^h(k+1) > I^h(i-k) \text{ for } k+1 \leq i \leq 2k-1.$$

Theorem 3.4.8 now follows from Lemma 3.3.10. ■

We next prove

THEOREM 3.4.9 $I^h(2k) > I^h(i)$ *for all* $i > 2k$.

Proof. By the Theorem 3.3.11, Theorem 3.4.9 holds for $i = 2k+1$. We prove the general case by induction on i.

By Corollary 3.4.3,

$$I^h(i,n) - I^h(2k,n) = \sum_{j=1}^{k} pq^{j-1}\Big(I^h(i-j,n-j) - I^h(2k,n-j)\Big) < 0$$

since each term is nonpositive and at most one term is zero by Theorem 3.4.8 and by induction. ■

COROLLARY 3.4.10 $I^h(3k) > I^h(3k+1)$.

Proof. Setting $i = 3k$ in Lemma 3.3.10. ■

We summarize the chief finding for the half-line case:

$$I^h(1) < I^h(2) < \ldots < I^h(k-1) < I^h(k+1) < I^h(i) < I^h(2k) < I^h(k)$$

for all $i > k+1$ and $i \neq 2k$.

3.5 THE COMBINATORIAL IMPORTANCE AND THE RARE-EVENT IMPORTANCE

When $p = 1/2$, the Birnbaum importance is called the *structure importance* by Barlow and Prochan [16]. We will rename it the *combinatorial importance* since it merely counts the number of success cases, reserving the term "structure importance" for general use. Formally, for fixed k, define the combinatorial importance

$$I^C(i) = 2\Big(R_L(i-1)R_L(n-i) - R_L(n)\Big).$$

If necessary, we use $I^C(i,n)$ to denote $I^C(i)$ with n specified. Lin, Kuo and Hwang [157] found an interesting correspondence between $I^C(i)$ and the *Fibonacci numbers* $f_{k,n}$ *of order* k. Define

$$f_{k,n} = \begin{cases} 0, & \text{if } 1 \le n \le k-1; \\ 1, & \text{if } n = k; \\ \sum_{i=1}^{k} f_{k,n-i}, & \text{if } n \ge k+1. \end{cases}$$

LEMMA 3.5.1 *$f_{k,n} = 2f_{k,n-1} - f_{k,n-k-1}$ for $n \ge k+2$.*

Proof. By the definition,

$$\begin{aligned} f_{k,n} &= f_{k,n-1} + (f_{k,n-2} + \ldots + f_{k,n-k}) \\ &= f_{k,n-1} + (f_{k,n-1} - f_{k,n-k-1}) \\ &= 2f_{k,n-1} - f_{k,n-k-1}. \end{aligned}$$

■

THEOREM 3.5.2 *For $p = 1/2$, $R_L(n) = (1/2)^n f_{k,n+k+1}$.*

Proof. It can be verified that

$$f_{k,n} = 2^{n-k-1} \quad \text{for } k+1 \le n \le 2k.$$

Thus for $0 \le n \le k-1$,

$$R_L(n) = 1 = (1/2)^n f_{k,n+k+1}.$$

We prove the general case by induction on n. Using Corollary 2.1.4,

$$\begin{aligned} R_L(n) &= \sum_{i=0}^{k-1} (1/2)^{i+1} R_L(n-i-1) \\ &= \sum_{i=0}^{k-1} (1/2)^{i+1} (1/2)^{n-i-1} f_{k,n-i+k} \\ &= (1/2)^n f_{k,n+k+1}. \end{aligned}$$

■

Thus $f_{k,n+k+1}$ can be interpreted as the number of working consecutive-k-out-of-n systems.

COROLLARY 3.5.3 $I(i,n) = (1/2)^{n-1}(2f_{k,i+k}f_{k,n-i+k-1} - f_{k,n+k+1})$.

Note that Ferguson [76] gave a closed form solution of $f_{k,n}$,

$$f_{k,n} = 2^{n-k-1} + \sum_{j=1}^{m}(-1)^j 2^{n-jk-k-j-1}\frac{(n-jk-k-1)!(n-jk-k+j)}{j!(n-jk-k-j)!}.$$

Recently, Chang and Hwang [45] introduced the notion of *rare-event importance* $I^R(i)$ which is defined to be the number of minimum pathsets containing i. The justification of studying $I^R(i)$ is that a conjecture of $I(i) > I(j)$ can be easily disproved by showing $I^R(i) < I^R(j)$. Also if $I^h(i) > I^h(j)$, then proving $I^R(i) > I^R(j)$ provides strong evidence for $I(i) > I(j)$ since the rare-event inequality covers the other end of p.

Represent n as $n = qk + r$, where $0 \le r < k$. Then q is the minimum number of working components for a pathset to exist. The rare-event importance $I^R(i)$ of component i is defined to be $|\mathrm{PS}_{i,q}(k,n)| \equiv \mathrm{ps}_{i,q}(k,n)$. Chang and Hwang proved

THEOREM 3.5.4 $\mathrm{ps}_q(k,n) = \binom{q+k-r-1}{k-r-1}$.

Proof. Theorem 3.5.4 is easily verified for $n \le k$. We prove the general case by induction on n.

Consider $n = qk + r > k$. Suppose $r > 0$. Then

$$n - 1 = qk + (r-1).$$

Clearly, every pathset of the n-line is a pathset of the $(n-1)$-line. The additional pathsets of the $(n-1)$-line are those which has its last working component at position $n-k$ (so that the n-line is not working). The number of such pathsets is of course $\mathrm{ps}_{q-1}(k, n-k-1)$. Therefore,

$$\begin{aligned}
\mathrm{ps}_q(k,n) &= \mathrm{ps}_q(k,n-1) - \mathrm{ps}_{q-1}(k,n-k-1) \\
&= \binom{q+k-(r-1)-1}{k-(r-1)-1} - \binom{(q-1)+k-(r-1)-1}{k-(r-1)-1} \\
&= \binom{q+k-r}{k-r} - \binom{q+k-r-1}{k-r} \\
&= \binom{q+k-r-1}{k-r-1}.
\end{aligned}$$

Next suppose $r = 0$. The last working component must occupy one of the last k positions, say m, and the first $m-1$ components must form a working line with $q-1$ working components. Therefore,

$$\begin{aligned}
\mathrm{ps}_q(k,n) &= \sum_{i=1}^{k} \mathrm{ps}_{q-1}(k,n-i) \\
&= \sum_{i=1}^{k} \binom{q-1+k-(k-i)-1}{k-(k-i)-1} \\
&= \sum_{i=1}^{k} \binom{q+i-2}{i-1} \\
&= \binom{q+k-1}{k-1}.
\end{aligned}$$

■

Represent $i = uk + v$, where $0 < v \le k$. Define $\binom{x}{y} = 0$ for $y < 0$ as usual.

THEOREM 3.5.5 $\mathrm{ps}_{i,q}(k,n) = \binom{u+k-v}{k-v}\binom{q-u+v-r-2}{v-r-1}$.

Proof. $\mathrm{PS}_{i,q}(k,n)$ consists of those pathsets with u working components in the first $i-1$ positions and $q-u-1$ working components in the last $n-i$ positions. In other words, it consists of the product of $\mathrm{PS}_u(k,i-1)$ and $\mathrm{PS}_{q-u-1}(k,n-i)$. Therefore,

$$\begin{aligned}
\mathrm{ps}_{i,q}(k,n) &= \mathrm{ps}_u(k,i-1)\mathrm{ps}_{q-u-1}(k,n-i) \\
&= \binom{u+k-(v-1)-1}{k-(v-1)-1}\binom{(q-u-1)+k-(k+r-v)-1}{k-(k+r-v)-1} \\
&= \binom{u+k-v}{k-v}\binom{q-u+v-r-2}{v-r-1},
\end{aligned}$$

by using Theorem 3.5.4 and noting the residue of $(n-i)/k$ is $k+r-v$.
■

LEMMA 3.5.6 *For* $k > v \ge r+1$, $\mathrm{ps}_{(u+1)k+v,q}(k,n) > \mathrm{ps}_{uk+v,q}(k,n)$ *if* $q > (u+1)(k-r-1)/(k-v)$.

Proof. By Theorem 3.5.5

$$\frac{\mathrm{ps}_{(u+1)k+v,q}(k,n)}{\mathrm{ps}_{uk+v,q}(k,n)} = \frac{(u+1+k-v)(q-u-1)}{(u+1)(q-u+v-r-2)}.$$

Hence $\mathrm{ps}_{(u+1)k+v,q}(k,n) > \mathrm{ps}_{uk+v,q}(k,n)$ if $q > (u+1)(k-r-1)/(k-v)$.
■

COROLLARY 3.5.7 *If* $v = r + 1 \neq k$, *then* $I^R(uk + v)$ *is nondecreasing in* u.

Proof. If $v = r + 1 \neq k$, then $q > (u + 1)(k - r - 1)/(k - v) = u + 1$. It implies, by Lemma 3.5.6, $\mathrm{ps}_{(u+1)k+v,q}(k, n) > \mathrm{ps}_{uk+v,q}(k, n)$. Hence, $I^R(uk + v)$ is nondecreasing in u. ■

THEOREM 3.5.8 $I^R(uk + v) \leq I^R((u + 1)k + v)$ *for* $1 \leq v \leq (k + 1)/2$ *and* $(u + 1)k + v \leq (n + 1)/2$.

Proof. If $v < r + 1$, then by Theorem 3.5.5, $I^R(uk + v) = 0$ for all u and Theorem 3.5.8 is trivially true. Hence we assume $v \geq r + 1$. Note that in the range of $(u + 1)k + v$ specified in Theorem 3.5.8,

$$qk + r \equiv n \geq 2\Big((u + 1)k + v\Big) - 1 = 2(u + 1)k + 2v - 1.$$

Since

$$2v - 1 > v - 1 \geq r,$$

necessarily, $q \geq 2u+3$. Finally, $v \leq (k+1)/2$ implies $(k-1)/(k-v) \leq 2$. Hence

$$\frac{(u + 1)(k - r - 1)}{k - v} \leq \frac{(u + 1)(k - 1)}{k - v} \leq 2(u + 1) < q.$$

Theorem 3.5.8 now follows from Lemma 3.5.6. ■

Although Theorem 3.5.8 deals with only the first half components, we can compare the second half by noting $I^R(i) = I^R(n + 1 - i)$.

The range of v in Theorem 3.5.8 cannot be pushed further in general. For example,

Example 3.1. $k = 4, n = 32, \mathrm{ps}_{15,8}(4, 32) = 60 < \mathrm{ps}_{11,8}(4, 32) = 63$.

Next, we compare the rare-event importance of position $uk + 1$ with the others.

THEOREM 3.5.9 $I^R(uk + 1) < I^R(uk)$ *for all* u *satisfying* $uk + 1 \leq (n + 1)/2$.

Proof. By Theorem 3.5.5,

$$\begin{aligned}\mathrm{ps}_{uk+1,q}(k, n) &= \binom{u + k - 1}{k - 1}\binom{q - u + 1 - r - 2}{1 - r - 1} \\ &= \begin{cases} \binom{u+k-1}{k-1}, & \text{if } r = 0; \\ 0, & \text{otherwise.} \end{cases}\end{aligned}$$

For $r = 0$, $uk + 1 \leq (n+1)/2$ implies $q \geq 2u + 1$.

$$\mathrm{ps}_{uk,q}(k,n) = \binom{q-u+1-r-1}{k-r-1} > \mathrm{ps}_{uk+1,q}(k,n).$$

Hence $I^R(uk+1) < I^R(uk)$ for all u. ■

Remark. Theorem 3.5.9 is not true for Birnbaum importance.

THEOREM 3.5.10 *$I^R(uk+1) \leq I^R(j)$ for all $uk+1 < j \leq (n+1)/2$.*

Proof. If $r \neq 0$, then by Theorem 3.5.5, $I^R(uk+1) = 0$ for all u and Theorem 3.5.10 is trivially true. Consider $r = 0$. First, we compare $I^R(uk+1)$ with $I^R(uk+v)$ for $2 \leq v \leq k$.

$$\frac{\mathrm{ps}_{uk+v,q}(k,n)}{\mathrm{ps}_{uk+1,q}(k,n)} =$$

$$\frac{(k-v+1)(k-v+2)\ldots(k-1)(q-u)(q-u+1)\ldots(q-u+v-2)}{(u+k-v+1)(u+k-v+2)\ldots(u+k-1)\cdot 1\cdot 2\ldots(v-1)}.$$

Since $q \geq 2u+1$, the ratio is increasing in q, we only need to consider $q = 2u+1$.

$$\frac{\mathrm{ps}_{uk+v,q}(k,n)}{\mathrm{ps}_{uk+1,q}(k,n)} =$$

$$\frac{(k-v+1)(k-v+2)\ldots(k-1)(u+1)(u+2)\ldots(u+v-2)}{(u+k-v+1)(u+k-v+2)\ldots(u+k-1)\cdot 1\cdot 2\ldots(v-1)}.$$

Since

$$(k-v+l)(u+l) \geq l(u+k-v+l) \text{ for } 1 \leq l \leq v-1,$$

$I^R(uk+1) < I^R(uk+v)$. By Theorem 3.5.8, $I^R(uk+1) \leq I^R((u+1)k+1)$. Hence $I^R(uk+1) < I^R(j)$ for all $j > uk+1$. ■

Remark. Theorem 3.5.10 is not true for Birnbaum importance.

Using I^R, we show that Theorem 3.4.6 cannot be extended to Birnbaum importance.

Example 3.2. For $k = 5$, $n = 25$, $\mathrm{ps}_{10,5}(5,25) = 35 < \mathrm{ps}_{13,5}(5,25) = 36$.

Example 3.3. For $k = 4$, $n = 16$, $\mathrm{ps}_{5,4}(4,16) = 4 < \mathrm{ps}_{6,4}(4,16) = 9 < \mathrm{ps}_{8,4}(4,16) = 10 < \mathrm{ps}_{7,4}(4,16) = 12$.

3.6 CONSECUTIVE-k G SYSTEM

A *consecutive-k G system* is the counterpart of a consecutive-k system (which will be referred to as the F system for contrast) by interchanging the notions of working and failed components, i.e., a system works if and only if some k consecutive components all work. Tong [244] first suggested such a system. As far as the computation of system reliability is concerned, formulas for the F systems also work for the G system by interchanging p_i and q_i for all i, and interchanging R with $\overline{R}$. Namely, what is computed as reliability for the F system is interpreted as the unreliability of the G system, after the switching of p_i and q_i in the formula.

However, the G system does bring out new problems in optimum sequencing and in component importance. This is because maximizing the reliability of the G system is equivalent to minimizing the reliability of the F system, which has not been studied before. Again, the notion of singularity plays a crucial role, but nonsingularity becomes undesirable.

Zuo and Kuo [270] gave a nonexistence result.

THEOREM 3.6.1 *There does not exist an invariant consecutive-k-out-of-n G line for* $2 \leq k < n/2$.

Proof. Choose x and y such that $0 < x < y < 1$. First consider the case:

(i) $p_{[1]} = p_{[2]} = \ldots = p_{[n-k-1]} = 0$, $p_{[n-k]} = p_{[n-k+1]} = y$, $p_{[n-k+2]} = p_{[n-k+3]} = \ldots = p_{[n]} = 1$.

Clearly, an optimal line must contain the subsequence $y, 1, \ldots, 1, y$ of length $k+1$. Let L_i denote the set of lines such that the subsequence of the $k+1$ best components start at node i. We consider two more cases which contradict each other.

(ii) $p_{[1]} = p_{[2]} = \ldots = p_{[n-k-1]} = x$, $p_{[n-k]} = p_{[n-k+1]} = y$, $p_{[n-k+2]} = p_{[n-k+3]} = \ldots = p_{[n]} = 1$.

From (i), it suffices to consider L_i, $i = 1, 2, \ldots, n-k$. If one y-node works, then all L_i works. But if both y-nodes are failed, then L_i works if and only if either the first $i-1$ nodes, or the last $n-k-i$ nodes, contain k consecutive working nodes. The probability is clearly maximized at $i = 1$ (or $i = n-k$). Thus an invariant line must be in L_1.

(iii) $p_{[1]} = p_{[2]} = \ldots = p_{[n-k-1]} = x \to 0$ and $p_{[n-k]} = p_{[n-k+1]} = \ldots = p_{[n]} = y \to 1$.

Again, we consider only L_i. If some k consecutive y-nodes work, then every L_i works. Otherwise, the dominant case is that one x node and its adjacent $k-1$ y-nodes all work. But L_1 has only one such event while L_2 has two. ∎

COROLLARY 3.6.2 *There does not exist an invariant consecutive-k-out-of-n G cycle for $2 \leq k < (n-1)/2$.*

Proof. Set $p_{[n]} = 1$ and reduce the n-cycle to the $(n-1)$-line. ■

Since each cutset for the F-system is a *pathset*, i.e., the system works if all components in the pathset work, we have

LEMMA 3.6.3 *If position i is criticality more important than position j, then $p_i \geq p_j$ in an optimal assignment of a G system.*

Zuo and Kuo first observed:

COROLLARY 3.6.4 *In an optimal assignment of a G system,*

$$p_1 \leq p_2 \leq \cdots \leq p_k \quad \textit{and} \quad p_n \leq p_{n-1} \leq \cdots \leq p_{n-k+1}.$$

The case of $n \leq 2k$ for the G line was studied by Kuo, Zhang and Zuo [150], while the case of $n \leq 2k+1$ for the G cycle by Zuo and Kuo [270]. However, both proofs are incomplete. Zuo and Kuo showed that interchanging any p_i and p_j in $\widetilde{C}_G$ (see Figure 3.6.1) will cause a decrease in reliability. However, this only proves that $\widetilde{C}_G$ cannot be improved by the local operation of interchanging two components. A similar problem exists for the proof of line.

Recently Jalali, Hawkes, Cui and Hwang [123] established the following lemma through using Corollary 3.6.4 and some local operations as interchanging two components, or shifting the assignment.

LEMMA 3.6.5 *In an optimal assignment of a G system, $p_1 = p_{[1]}$, $p_k = p_{[n-1]}$, $p_{k+1} = p_{[n]}$ and $p_n = p_{[2]}$.*

Using this lemma, they proved the line case.

THEOREM 3.6.6 *The unique invariant consecutive-k-out-of-n G line for $n = 2k$ is $\alpha = (p_{[1]}, p_{[3]}, p_{[5]}, \ldots, p_{[6]}, p_{[4]}, p_{[2]})$.*

Proof. The theorem is trivially true for $k = 1$ and follows from Lemma 3.6.5 for $k = 2$. We prove the general case by induction on k.

Suppose to the contrary that $\beta = (p'_1, p'_2, \ldots, p'_n) \neq \alpha$ is an optimal line. By Lemma 3.6.5, we have $p'_1 = p_{[1]}$, $p'_k = p_{[n-1]}$, $p'_{k+1} = p_{[n]}$ and $p'_n = p_{[2]}$. Let $F_\alpha, S_\alpha, F_\beta, S_\beta$ denote the product of reliabilities for the first or second half of α or β, respectively. In the proof of Lemma 3.6.5, it was proved

$$F_x = \left(\prod_{i=1}^{k-1} p_i\right) p_k < \left(\prod_{i=k+2}^{n} p_i\right) p_{k+1} = S_x \quad \text{for } x \in \{\alpha, \beta\}.$$

Let σ be a sequence of the component reliabilities, and let $t(\sigma) \equiv t_\sigma$ be a subsequence obtained from σ by removing the smallest and the largest reliability. We will write α by α_k if the parameter k needs to be noted. Then $t(\alpha_k) = \alpha_{k-1}$. Since β is optimal,

$$0 \geq R(\alpha) - R(\beta) = q_{k-1}(F_\alpha - F_\beta) + p_{k-1}(R(t_\alpha) - R(t_\beta)). \quad (3.6.1)$$

By the induction hypothesis, t_α is optimal. Hence the second term is nonnegative. It follows $F_\alpha \leq F_\beta$. Thus we have

$$0 < F_\alpha \leq F_\beta < S_\beta \leq S_\alpha < 1.$$

Consequently,

$$S_\alpha - S_\beta = S_\alpha - F_\alpha S_\alpha / F_\beta = (F_\beta - F_\alpha) S_\alpha / F_\beta > F_\beta - F_\alpha. \quad (3.6.2)$$

We can expand $R(t_\alpha) - R(t_\beta)$ in (3.6.1) in the same way we expand $R(\alpha) - R(\beta)$. Thus

$$\begin{aligned} 0 &\geq R(\alpha) - R(\beta) \\ &= q_{k+1}(F_\alpha - F_\beta) + \\ &\quad p_{k+1}\left(q_k\left(\prod_{i=k+2}^{n} p_i - \prod_{i=k+2}^{n} p_i'\right) + p_k\Big(R(t(t_\alpha)) - R(t(t_\beta))\Big)\right) \\ &\geq q_{k+1}(F_\alpha - F_\beta) + q_k(S_\alpha - S_\beta), \end{aligned}$$

since $t(t_\alpha)$ is optimal by the induction hypothesis. But

$$q_{k+1} < q_k;$$

hence

$$F_\beta - F_\alpha > S_\alpha - F_\beta,$$

a contradiction to (3.6.2). Therefore β is not optimal. ■

COROLLARY 3.6.7 *For $n \leq 2k$, an invariant line is $(p_{[1]}, p_{[3]}, p_{[5]}, \ldots B \ldots, p_{[6]}, p_{[4]}, p_{[2]})$ where B is a center block of $2k - n$ largest reliabilities in any arrangement.*

Proof. Any working state must have every component in B working. Therefore B should consist of the largest reliabilities. Furthermore, the system L works if and only if all components in B work and some $n - k$ components consecutive in $L \backslash B$ all work. Therefore, the components not in B should be arranged according to Theorem 3.6.6 with $n' = 2k'$ and $k' = n - k$. ■

Since the cycle case is still open, we state it as a conjecture.

CONJECTURE 3.6.8 *The unique invariant consecutive-k-out-of-n G cycle for* $n \leq 2k+1$ *is* $\tilde{C}_G = (p_{[1]}, p_{[3]}, p_{[5]}, \ldots, p_{[6]}, p_{[4]}, p_{[2]})$, *where the two ends are considered adjacent (see Figure 3.6.1).*

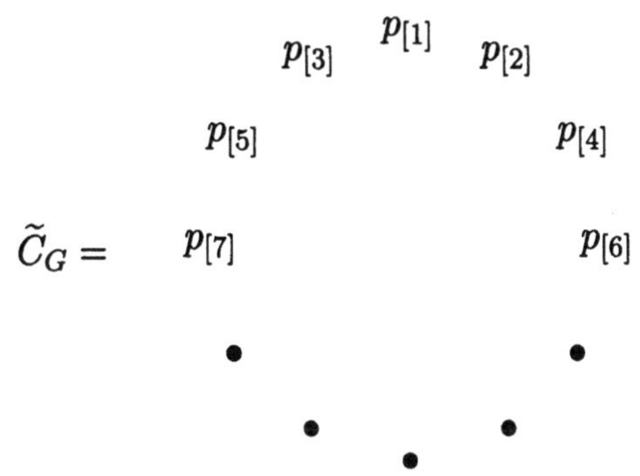

Figure 3.6.1. The unique invariant consecutive-*k*-out-of-*n* G cycle.

THEOREM 3.6.9 *Conjecture 3.6.8 holds for* $n = k+2$.

Proof. The cycle works if and only if either there exists at most one failed component, or there exist exactly two but they are adjacent. Since sequencing only affects the last case, we want to maximize

$$f(C) = \sum_{i=1}^{n} q_i q_{i+1} \prod_{j \neq i, i+1} p_j = \prod_{j=1}^{n} p_j \sum_{i=1}^{n} \rho_i \rho_{i+1},$$

where $\rho_i = q_i/p_i$. Suppose a nonsingular index exists, i.e., $p_i > p_j$ but $p_{i+1} < p_{j-1}$. Then $f(C)$ will increase by reversing the subsequence $(p_{i+1}, p_{i+2}, \ldots, p_{j-1})$ (call the new cycle C') since

$$\begin{aligned} f(C') - f(C) &= \prod_{j=1}^{n} p_j (\rho_i \rho_{j-1} + \rho_j \rho_{i+1} - \rho_i \rho_{i+1} - \rho_j \rho_{j-1}) \\ &= \prod_{j=1}^{n} p_j (\rho_i - \rho_j)(\rho_{j-1} - \rho_{i+1}) > 0 \ \ (\text{noting } \rho_i \downarrow p_i). \end{aligned}$$

It is easily verified that the cycle shown in Figure 3.6.1 is the unique cycle without a nonsingularity index. ■

Kuo, Zhang and Zuo [150] obtained

THEOREM 3.6.10 *For the G system,*

$$I_L(i) = \Big(\overline{R}_L(1, i-1)\overline{R}_L(i+1, n) - \overline{R}_L(1, n)\Big)/p_i,$$
$$I_C(i) = \Big(\overline{R}_L(p_{i+1}, \ldots, p_{i-1}) - \overline{R}_C(1, n)\Big)/p_i.$$

Proof. For $x \in \{L, C\}$,

$$\begin{aligned} I_x(i) &= R_x(p_1, \ldots, p_{i-1}, 1, p_{i+1}, \ldots, p_n) - \\ &\quad R_x(p_1, \ldots, p_{i-1}, 0, p_{i+1}, \ldots, p_n) \\ &= \Big(R_x(1, n) - q_i R_x(p_1, \ldots, p_{i-1}, 0, p_{i+1}, \ldots, p_n)\Big)/p_i - \\ &\quad R_x(p_1, \ldots, p_{i-1}, 0, p_{i+1}, \ldots, p_n) \\ &= \Big(R_x(1, n) - R_L(p_1, \ldots, p_{i-1}, 0, p_{i+1}, \ldots, p_n)\Big)/p_i \\ &= \Big(\overline{R}_x(p_1, \ldots, p_{i-1}, 0, p_{i+1}, \ldots, p_n) - \overline{R}_x(1, n)\Big)/p_i. \end{aligned}$$

The formula for $I_L(i)$ then follows from that

$$\overline{R}_L(p_1, \ldots, p_{i-1}, 0, p_{i+1}, \ldots, p_n) = \overline{R}_L(1, i-1)\overline{R}_L(i+1, n)$$

and the formula for $I_C(i)$ from that

$$\overline{R}_C(p_1, \ldots, p_{i-1}, 0, p_{i+1}, \ldots, p_n) = \overline{R}_L(p_{i+1}, \ldots, p_{i-1}).$$

■

COROLLARY 3.6.11 *For the IID model,* $I_L(i) = \Big(\overline{R}_L(i-1)\overline{R}_L(n-i) - \overline{R}_L(n)\Big)/p$ *and* $I_C(i) = \Big(\overline{R}_L(n-1) - \overline{R}_C(n)\Big)/p$.

To interplay between the F system and the G system, we need to add F or G as well as p, the parameter of the component reliability, into the notation. Zuo and Kuo [270] proved the following surprising result.

THEOREM 3.6.12 *Consider a consecutive-2 G line under IID model. Then the ranking of* $\{I_L(i; G)\}$ *is same as* $\{I_L(i; F)\}$.

Proof. $I_L(i, G) = \Big(\overline{R}_L(i-1; G, p)\overline{R}_L(n-i; G, p) - \overline{R}_L(n; G, p)\Big)/p = \Big(R_L(i-1; F, q)R_L(n-i; F, q) - R_L(n; F, q)\Big)/p = qI_L(i; F, q)/p$, which is in the same form as $I_L(i; F)$ except the swapping of p and q. But the argument used in proving Theorem 3.3.4 is independent of the quantity of p. Hence, the conclusion of the theorem holds here too. ■

Zuo [266] claimed that $I_L(i;G) = I_L(i;F)$ for all k. Hence they have the same ordering. First of all, $I_L(i;G) \neq I_L(i;F)$ since one can be transformed to the other only by switching q and p. Second, Hwang, Cui, Chang and Lin [110] gave a counterexample that for $n = 7$ and $k = 3$, $I_L(2;F) > I_L(4;F)$ for $p = 0.2$, but $I_L(2;F) < I_L(4;F)$ for $p = 0.8$. It follows that for $p = 0.8$, $I_L(2;G) > I_L(4;G)$.

The heuristics we mentioned in Section 3.3 for the F system apply to the G system too.

Chapter 4

THE LIFETIME DISTRIBUTION

In this chapter we assume that component i has a *lifetime* distribution, i.e., the probability that component i is still working at time t is $p_i(t)$. In the IND model $p_i(t)$ and $p_j(t)$ are independent for $i \neq j$. In the IID model $p_i(t) = p(t)$ for all i. We also use $F(t)$ for $q(t)$ and $1 - F(t)$ for $p(t)$. Let $R(t)$ denote the reliability of the system at time t, i.e., the probability that the system works from time 0 to time t. Since no repair is involved, this dynamic probability is same as the static probability that the system works at time t which can be computed using the formulas of Chapter 2 by simply replacing p_i with $p_i(t)$. However, there exist problems unique to the dynamic process, which will be studied in this chapter.

4.1 MEAN TIME TO FAILURE

Papastavridis and Hadjichristos [195] studied the mean time to failure (see corrections by Sasaki, Nakai and Yuge [227] and Sasaki [226]).

THEOREM 4.1.1 *The mean time to failure of consecutive-k linear and circular systems are*

$$\begin{aligned}
\mu_L &= \sum_{m=0}^{(k+2)n} (-1)^m \int_0^\infty p(t)^m dt \\
&\quad \sum_{i=m-n}^{m} \left(\binom{n-ki}{i} \binom{ki}{m-i} - \binom{n-ki-k}{i} \binom{ki+k}{m-i} \right), \\
\mu_C &= -\sum_{i=0}^{n} \binom{n}{i} (-1)^i \int_0^\infty p(t)^i dt +
\end{aligned}$$

$$\sum_{m=0}^{(k+2)n} \sum_{i=m-n}^{n} \left((-1)^m \int_0^\infty p(t)^m dt \binom{n-ki}{i} \binom{ki}{m-i} + (-1)^{m+1} \int_0^\infty p(t)^{m+1} dt \binom{n-ki-k-1}{i} \binom{ki+k}{m-i} \right).$$

Proof. Note that $\mu_L = \int_0^\infty R_L(t)dt$ and use Corollary 2.3.6 for $R_L(t)$. By expanding $q^x = (1-p)^x = \sum_{j=0}^{x} \binom{x}{j}(-p)^j$ in Theorem 2.3.8 and setting $m = i + j$ for changing variables, we have

$$\begin{aligned} R_C(n) &= -\sum_{i=0}^{n} \binom{n}{i}(-p)^i + \sum_{m=0}^{(k+2)n} \sum_{i=m-n}^{n} \left((-p)^m \binom{n-ki}{i} \right. \\ & \left. \binom{ki}{m-i} + (-p)^{m+1} \binom{n-ki-k-1}{i} \binom{ki+k}{m-i} \right). \end{aligned}$$

The formula for μ_C then follows by taking $\mu_C = \int_0^\infty R_C(t)dt$. ∎

COROLLARY 4.1.2 *Suppose $p(t)$ is the Weibull distribution $W(\alpha, \beta)$. Then*

$$\begin{aligned} \mu_L &= \alpha\Gamma(1+1/\beta) \sum_{m=0}^{(k+2)n} (-1)^m m^{-1/\beta} \\ & \sum_{i=m-n}^{m} \left(\binom{n-ki}{i} \binom{ki}{m-i} - \binom{n-ki-k}{i} \binom{ki+k}{m-i} \right), \\ \mu_C &= \alpha\Gamma(1+1/\beta) \left(\sum_{m=0}^{(k+2)n} (-1)^m \sum_{i=m-n}^{m} \left(m^{-1/\beta} \binom{n-ki}{i} \binom{ki}{m-i} - \right. \right. \\ & \left. \left. k(m+1)^{-1/\beta} \binom{n-ki-k-1}{i} \binom{ki+k}{m-i} \right) - \sum_{i=0}^{n} \binom{n}{i} (-1)^i i^{-1/\beta} \right). \end{aligned}$$

Proof. For the Weibull distribution, $p(t) = \exp(-t^\beta/\alpha^\beta)$. The corollary then follows from Theorem 4.1.1 and the Gamma formula

$$\int_0^\infty \exp(-xt^\beta/\alpha^\beta)dt = \alpha\Gamma(1+1/\beta)x^{-1/\beta}.$$

∎

They also indicated that higher moments can be obtained with a similar method.

Bolinger and Savia [31] proposed a different approach. A sequence of components is called a *cut-sequence* if their collective failures induce the system failure. It is a *minimal cut-sequence* if no proper subsequence is a cut-sequence. Let n_d denote the number of minimal cut-sequences of size d. Hwang [107] proved

LEMMA 4.1.3 *For* $n \geq d \geq k$,

$$n_d = (n-d+1)(d-1)! \sum_{i=0}^{\lfloor n/k \rfloor} (-1)^i \binom{n-d+2}{i} \binom{n-ki}{n-d+1} - d! \sum_{i=0}^{\lfloor n/k \rfloor} (-1)^i \binom{n-d+1}{i} \binom{n-ki}{n-d}.$$

Proof. Recall that $N(d,n,k)$ is the number of ways of selecting d failed components among an n-line not containing a cutset. There are $d!$ ways of arranging the d selected components into a sequence. Therefore, the number of $(m-1)$-sequences not being cut-sequence is $(d-1)!N(d-1,n,k)$, and there are $n-d+1$ ways of selecting the dth failed components. After subtracting $d!N(d,n,k)$, the number of d sequences not being cut-sequences, we obtain the number of d-sequences which are minimal cut-sequences. Lemma 4.1.3 follows by substituting the closed form expression of $N(d,n,k)$ given in Theorem 2.3.3. ∎

THEOREM 4.1.4

$$R_L(t) = \sum_{d=k}^{n} \left(1 - \frac{n_d}{n(n-1)\dots(n-d+1)}\right) \binom{n}{d} (1-F(t))^{n-d} F^d(t).$$

COROLLARY 4.1.5

$$\mu_L = \int_0^\infty \sum_{d=k}^{n} \left(1 - \frac{n_d}{n(n-1)\dots(n-d+1)}\right) \binom{n}{d} (1-F(t))^{n-d} F^d(t) dt.$$

Papastavridis [190] studied the distribution of the number of failed components at the time of system failure. Let D be the random variable representing the number of failed components.

THEOREM 4.1.6

$$\Pr(D=d) = \sum_{m=k}^{\min\{d,2k-1\}} \sum_{i=1}^{n-m-1} \sum_{j=0}^{\max\{0,i-2\}} \frac{2k-n}{n-d+1} \frac{1}{\binom{n}{d-1}} N(j,i-2,k) \\ N(d-m-j, n-i-m, k),$$

Proof. Consider the $n!$ permutations of components in order of their failure times. Then $\Pr(D = d)$=number of permutations where the first system failure occurs at the dth component failure/$n!$.

We now compute the numerator. There must exist exactly one sequence $I = (i, i+1, \ldots, i+m-1)$ of m consecutive failed components for some $k \leq m \leq \min\{d, 2k-1\}$, where i can be chosen from 1 to $n-m+1$. Furthermore, the $d-m$ other failed components split into the two intervals $(1, 2, \ldots, i-2)$ and $(i+m+1, i+m+2, \ldots, n)$ (by definition of I, components $i-1$ and $i+m$ must be working) which must both work. Let j be the number of failed components, $0 \leq j \leq \max\{0, i-2\}$, in the first interval. Then the number of ways of having both intervals working is

$$N(j, i-2, k)N(d-m-j, n-i-m, k).$$

Finally, the dth failed component f must be in I, and can be any of the $2k-m$ choices such that $I - \{f\}$ does not contain k consecutive failed components. But otherwise, the first $d-1$ failed components can appear in any order and so can the last $n-d$ failed components. Therefore, the numerator is

$$\sum_{m=k}^{\min\{d,2k-1\}} \sum_{i=1}^{n-m-1} \sum_{j=0}^{\max\{0.i-2\}} (2k-m)(d-1)!(n-d)!N(j, i-2, k)$$
$$N(d-m-j, n-i-m, k).$$

Theorem 4.1.6 follows immediately. ∎

COROLLARY 4.1.7

$$E(D) = \sum_{d=0}^{n} \sum_{m=k}^{\min\{d,2k-1\}} \sum_{i=1}^{n-m-1} \sum_{j=0}^{\max\{0,i-2\}} \frac{2k-m}{\binom{n}{d}} N(j, i-2, k)$$
$$N(d-m-j, n-i-m, k).$$

If the failure time t is also given, then Theorem 4.1.6 can be used.

THEOREM 4.1.8

$$\Pr(D = d|t) = \frac{n_d / \prod_{j=0}^{d-1}(n-j)\binom{n}{d}(1-F(t))^{n-d}F^d(t)}{\sum_{m=k}^{n}\left(n_m \prod_{j=0}^{m-1}(n-j)\right)\binom{n}{m}(1-F(t))^{n-m}F^m(t)}.$$

Proof. Use the Bayes theorem, the result can easily be gotten. ∎

COROLLARY 4.1.9

$$E(D) = \frac{\sum_{d=0}^{n} \left(dn_d / \prod_{j=0}^{d-1} (n-j) \binom{n}{d} (1-F(t))^{n-d} F^d(t) \right)}{\sum_{m=k}^{n} n_m / \left(\prod_{j=0}^{m-1} (n-j) \right) \binom{n}{m} (1-F(t))^{n-m} F^m(t)}.$$

4.2 ESTIMATION OF PARAMETERS

Aki and Hirano [3] studied the estimation of parameters for consecutive systems. They considered three distributions: the exponential, the Weibull and the Pareto distributions as component lifetime, respectively. They used the moment estimation method and the system lifetime data to give the estimators under the IID assumption.

Let $T_1, T_2, \ldots, T_m$ be an IID sample of lifetimes of a consecutive system. Suppose that the system component lifetimes are IID random variables $X_1, X_2, \ldots, X_n$ with order statistics $X_{[1]} \le X_{[2]} \le \ldots \le X_{[n]}$. By using the order statistics theory, it is well-known that the pdf of $X_{[i]}$ is

$$f_{[i]}(t) = \frac{n!}{(i-1)!(n-i)!} F^{i-1}(t)(1-F(t))^{n-i} f(t).$$

From Theorem 4.1.4, the pdf of T_1 can be written as

$$f_T(t) = f_T(t, \mu) = \sum_{i=1}^{n} \omega_i f_{[i]}(t),$$

where $\omega_i = 1 - \frac{n_i}{n(n-1)\ldots(n-i+1)}$.

(i) Exponential distribution

For the exponential distribution, the cdf of X_i is $F(x) = 1 - \exp(-\mu t)$, where $\mu > 0$. Then

$$f_{[i]}(t) = \frac{n!}{(i-1)!(n-i)!} (1 - \exp(-\mu t))^{i-1} \mu \exp(-\mu(n-i+1)t).$$

Thus

$$E(X_{[i]}) = \int_0^{\infty} t f_{[i]}(t) dt = \frac{1}{\mu} \sum_{j=n-i+1}^{n} \frac{1}{j},$$

$$\mathrm{Var}(X_{[i]}) = \frac{1}{\mu^2} \sum_{j=n-i+1}^{n} \frac{1}{j^2},$$

$$E(T_1) = \int_0^{\infty} t f(t) dt = \frac{A}{\mu},$$

$$\operatorname{Var}(T_1) = \int_0^\infty \Big(t - E(T_1)\Big)^2 f(t)dt = \frac{B}{\mu^2},$$

where

$$A = \sum_{i=1}^n \omega_i \Big(\sum_{j=n-i+1}^n \frac{1}{j} \Big),$$

$$B = \sum_{i=1}^n \omega_i \Big(\sum_{j=n-i+1}^n \frac{1}{j^2} + \Big(\sum_{j=n-i+1}^n \frac{1}{j} \Big)^2 \Big) - A^2.$$

Using the moment estimation method, we have

$$\frac{A}{\hat{\mu}_{ME}} = \frac{1}{m} \sum_{i=1}^m T_i \stackrel{\text{def}}{=} \overline{T}, \text{ i.e, } \hat{\mu}_{ME} = A/\overline{T},$$

where $\hat{\mu}_{ME}$ is the moment estimator of μ.

THEOREM 4.2.1 *$\sqrt{m}(\hat{\mu}_{ME} - \mu)$ converges in distribution to normal distribution with mean 0 and variance $\frac{B}{A^2}\mu^2$ as m tends to infinity.*

Proof. It follows from the central limit theorem that $\sqrt{m}(\overline{T} - h(\mu)) \longrightarrow N(0, \sigma^2(\mu))$, where $h(\mu) = \frac{A}{\mu}$ and $\sigma^2(\mu) = \operatorname{Var}(T_1) = \frac{B}{\mu^2}$. If $g(x)$ is a real valued function having a nonzero differential at $x = h(\mu)$, then $\sqrt{m}(g(\overline{T}) - g(h(\mu)))$ converges in distribution to $N(0, \sigma^2(\mu)(g'(h(\mu)))^2)$ (cf. e.g. Serfling [230]). By setting $g = h^{-1}$, the theorem is proved. ■

(ii) Weibull distribution

For the Weibull distribution, the cdf is $F(t) = 1 - \exp\{-(\frac{t}{\alpha})^\beta\}$. Then $f(t) = \frac{\beta}{\alpha}(\frac{t}{\alpha})^{\beta-1}\exp\{-(\frac{t}{\alpha})^\beta\}$. By the same method, the pdf of $X_{[i]}$ can be written as

$$f_{[i]}(t) = \frac{n!\beta t^{\beta-1}}{(i-1)!(n-i)!\alpha^\beta} \sum_{j=0}^{i-1} \binom{i-1}{j} (-1)^j \exp\Big(-(n-i+1+j)(t/\alpha)^\beta\Big).$$

The sth moment of the ith order statistic is

$$E(X_{[i]}^s) = \frac{n!\alpha^s}{(i-1)!(n-i)!} \Gamma(1+\frac{s}{\beta}) \sum_{j=0}^{i-1} \binom{i-1}{j} (-1)^j (n-i+1+j)^{-(1+\frac{s}{\beta})}.$$

Then

$$E(T_1) = \sum_{i=1}^n \omega_i \int_0^\infty t f_{[i]}(t)dt = \alpha A(\beta, 1),$$

$$E(T_1^2) = \sum_{i=1}^{n} \omega_i \int_0^\infty t^2 f_{[i]}(t)dt = \alpha^2 A(\beta, 2),$$

where

$$A(\beta, s) = \sum_{i=1}^{n} \frac{\omega_i n!}{(i-1)!(n-i)!}\Gamma(1+\frac{s}{\beta})\sum_{j=0}^{i-1}\binom{i-1}{j}(-1)^j(n-i+1+j)^{-(1+\frac{s}{\beta})}$$

for $s = 1, 2, \ldots$.

Let $M_1 = \frac{1}{m}\sum_{i=1}^{m} T_i$ and $M_2 = \frac{1}{m}\sum_{i=1}^{m} T_i^2$. Then, the solutions $\hat{\alpha}_{ME}$ and $\hat{\beta}_{ME}$ of the following equations are the moment estimators of α and β, respectively.

$$\begin{cases} \alpha A(\beta, 1) = M_1, \\ \alpha^2 A(\beta, 2) = M_2. \end{cases}$$

(iii) Pareto distribution

For the Pareto distribution, the cdf is $F(t) = 1 - (\frac{\alpha}{t})^{\beta-1}$, where $\alpha > 0$ and $\beta > 1$. Then $f(t) = \frac{1}{\alpha}\frac{\beta-1}{(t/\alpha)^\beta}$ for $t > \alpha$. The pdf of $X_{[i]}$ can be written as

$$f_{[i]}(t) = \frac{n!}{(i-1)!(n-i)!}\left(1 - (\alpha/t)^{\beta-1}\right)^{i-1}\frac{\beta-1}{\alpha}(\alpha/t)^{(\beta-1)(n-i)+\beta}.$$

The sth moment of the ith order statistic is

$$E(X_{[i]}^s) = \frac{n!\alpha^s}{(n-i)!}\frac{\Gamma(n-i-\frac{s}{\beta-1}+1)}{\Gamma(n-\frac{s}{\beta-1}+1)}.$$

Then

$$E(T_1^s) = (\beta-1)\alpha^s \sum_{j=1}^{n}\sum_{i=0}^{j-1} \frac{r_{jk}}{(j-i-1)!i!}(-1)^{(j-i-1)}\frac{1}{(\beta-1)(n-i)-s}.$$

By solving the following equations for α and β, we obtain the moment estimators $\hat{\alpha}_{ME}$ and $\hat{\beta}_{ME}$:

$$\begin{cases} E(T_1^1) = M_1, \\ E(T_1^2) = M_2. \end{cases}$$

On the other hand, if we have the component lifetime data, the parameters in component lifetime distribution can be estimated by the moment estimation method or the maximum likelihood estimation method etc. Then we use the reliability formula of consecutive k-out-of-n system to estimate the system reliability.

4.3 INCREASING FAILURE RATE PRESERVATION

Let $F(t)$ denote the failure time distribution, i.e., the probability that the object is failed at time t. Suppose $F(t)$ is differential with pdf $f(t)$. $\lambda(t) = \frac{f(t)}{1-F(t)}$ is called the *failure rate* of the object. We say the object has *increasing failure rate* (IFR) if $\lambda(t)$ is increasing (throughout this chapter the word "increasing" means "non-decreasing" except specially stated). Derman, Lieberman and Ross [68] raised the question whether a consecutive-k system preserves the IFR, i.e., whether its lifetime is IFR whenever the components have statistically independent and identically distributed IFR lifetimes. They gave the following result: $C(n,2)$ preserves IFR for all $n \geq 2$. They also showed that the result does not hold in general by giving counterexamples $C(6,3)$ and $L(3,2)$. Unfortunately, the proof for $C(n,2)$ was based on an incorrect assumption that in a working system i failed components had $2i$ "distinct" neighbor components (missing the fact that a component can be sandwiched between two failed components), as noted by Hwang and Yao [120], who presented a basic theory for IFR preservation as shown in this section.

Consider a system $S(n,k)$ ($S = L$ or C if the system is linear or circular). Let T be the lifetime of $S(n,k)$ and let $F_T(t)$ and $\lambda_T(t)$ denote the cdf and the failure rate of T. Then

$$\lambda_T(t) = F_T'(t)/(1 - F_T(t)).$$

Let $N_S(i,n,k)$ ($S = L$ or C if restricted to linear or circular system) denote the number of ways of arranging i failed components and $(n-i)$ working components into a working system, i.e., without k consecutive components all fail. Let $\sum_x$ denote the summation over all nonnegative integers x.

THEOREM 4.3.1 $\lambda_T(t) =$

$$\frac{\lambda(t)\sum_i \Big((n-i)N_S(i,n,k) - (i+1)N_S(i+1,n,k)\Big)F(t)^i(1-F(t))^{n-i}}{\sum_i N_S(i,n,k)F(t)^i(1-F(t))^{n-i}}.$$

Proof. By the definition of $N_S(i,n,k)$, we have

$$\begin{aligned} 1 - F_T(t) &= \sum_i N_S(i,n,k)F(t)^i(1-F(t))^{n-i}, \\ F_T'(t) &= -\frac{d}{dt}\sum_i N_S(i,n,k)F(t)^i(1-F(t))^{n-i} \end{aligned}$$

$$= \sum_i N_S(i,n,k)\Big((n-i)F(t)^i(1-F(t))^{n-i-1}$$
$$-iF(t)^{i-1}(1-F(t))^{n-i}\Big)F'(t)$$
$$= \sum_i \Big((n-i)N_S(i,n,k) - (i+1)N_S(i+1,n,k)\Big)$$
$$F(t)^i(1-F(t))^{n-i-1}F'(t)$$
$$= \sum_i \Big((n-i)N_S(i,n,k) - (i+1)N_S(i+1,n,k)\Big)$$
$$F(t)^i(1-F(t))^{n-i}\lambda(t).$$

Theorem 4.3.1 then follows. ∎

From now on we assume $1 < k < n$. Let

$$n = kq + r,\ 0 \le r < k \text{ for } L(n,k); \quad \text{and}$$
$$n = kq - r,\ 0 \le r < k \text{ for } C(n,k).$$

Thus, q is the minimum number of working components such that the system still works. In other words, $N_S(i,n,k) = 0$ for $i > n-q$ and $N_S(i,n,k) > 0$ for $i \le n-q$. We now define

$$G_S(p) = \frac{\sum_{i=0}^{n-q}\Big((n-i)N_S(i,n,k) - (i+1)N_S(i+1,n,k)\Big)p^i(1-p)^{n-q-i}}{\sum_{i=0}^{n-q} N_S(i,n,k)p^i(1-p)^{n-q-i}}.$$

Note that $\lambda_T(t) = \lambda(t)G_S(F(t))$ by Theorem 4.3.1.

THEOREM 4.3.2 *$S(n,k)$ preserves IFR if and only if $G_S(p)$ is increasing in $0 \le p < 1$.*

Proof. The "if" part is obvious. To prove the "only if" part, first set $\lambda(t) = 1$. Then $\lambda_T(t) = \lambda(t)G_S(F(t))$. Since $F(t)$ is increasing in t, $\lambda_T(t)$ increasing in t implies $G_S(p)$ increasing in t. The theorem follows by observing that $G_S(F(t))$ is independent of $\lambda(t)$. ∎

Define $\Delta = N_S(n-q-1,n,k) - (n-q)N_S(n-q,n,k)$.

THEOREM 4.3.3 *$S(n,k)$ does not preserve IFR if $\Delta > 0$.*

Proof. Note that $N_S(n-q,n,k) > 0$. So $G_S(p)$ is well-defined at $p = 1$.

$$G'_S(1) = \frac{N_S(n-q,n,k)\Big((n-q)qN_S(n-q,n,k) - (q+1) N_S(n-q-1,n,k+(n-q)N_S(n-q,n,k)\Big)}{N_S(n-q,n,k)^2} -$$

$$\frac{qN_S(n-q,n,k)\Big((n-q)N_S(n-q,n,k)-N_S(n-q-1,n,k)\Big)}{N_S(n-q,n,k)^2}$$

$$= -\frac{\Delta}{N_S(n-q,n,k)} < 0 \quad \text{if } \Delta > 0.$$

By continuity $G_S(p)$ strictly decreases in p when p is close to unity. Theorem 4.3.3 then follows from Theorem 4.3.2. ■

For a given $S(n,k)$, let w denote a subset of i components such that $S(n,k)$ remains working if every component in w fails but all other components work. Let W_i be the set of all such w. Clearly, $|W_i| = N_s(i,n,k)$. Let $U(w) = \{u : u$ is a component such that $u \cup w$ contains k consecutive components$\}$.

THEOREM 4.3.4

$$\lambda_T(t) = \frac{\lambda(t)\sum_i \sum_{w\in W_i} |U(w)|F(t)^i(1-F(t))^{n-i}}{\sum_i \sum_{w\in W_i} F(t)^i(1-F(t))^{n-i}}.$$

Proof. Let D denote the set of components which have failed before time t. Then

$$\begin{aligned}
\lambda_T(t)dt &= \Pr(t < T < t+dt | T > 0) \\
&= \sum_i \sum_{w\in W_i} \Pr(D = w | T > t)\Pr(t < Tt + dt | D = w) \\
&= \frac{\sum_i \sum_{w\in W_i} \Pr(D = w)|U(w)|\lambda(t)dt}{\Pr(T > t)} \\
&= \frac{\lambda(t)dt \sum_i \sum_{w\in W_i} |U(w)|F(t)^i(1-F(t))^{n-i}}{\sum_i \sum_{w\in W_i} F(t)^i(1-F(t))^{n-i}}
\end{aligned}$$

Theorem 4.3.4 then follows. ■

COROLLARY 4.3.5 *$S(n,k)$ dose not preserve IFR if $|U(w)| \geq q$ for every $w \in W_{n-q-1}$ and the strict inequality holds for at least one such w.*

Proof. By Theorems 4.3.1 and 4.3.4,

$$\begin{aligned}
\sum_{w\in W_{n-q-1}} |U(w)| &= (q+1)N_S(n-q-1,n,k) - (n-q)N_S(n-q,n,k) \\
&> \sum_{w\in W_{n-q-1}} q = qN_S(n-q-1,n,k),
\end{aligned}$$

which implies $N_S(n-q-1,n,k) > (n-q)N_S(n-q,n,k)$. The corollary then follows from Theorem 4.3.3. ■

This corollary transforms the IFR problem into a combinatorial problem which is different from Theorem 4.3.3.

THEOREM 4.3.6 *If $k \geq 2$, then $L(kq+k-1,k)$ for $q \geq 1$ and $C(kq,k)$ for $q \geq 2$ do not preserve IFR except for $C(4,2)$.*

THEOREM 4.3.7 *$L(n,k)$ preserves IFR for all (n,k) satisfying either $n = 2k \geq 6$ or $(3k-1)/2 \geq n \geq k \geq 3$.*

COROLLARY 4.3.8 *For every integer $d \geq 1, L(n,n-d)$ preserves IFR for all $n \geq 3d+1$.*

COROLLARY 4.3.9 *$L(n,k)$ does not preserve IFR for all $(3k-1)/2 < n < 2k$.*

THEOREM 4.3.10 *There exists $k_0 \geq 2$ such that $C(n,k)$ preserves IFR for all $k_0 \leq k < n \leq 1.02k$*

COROLLARY 4.3.11 *For every integer $d \geq 1$, there exists n_d such that $C(n,n-d)$ preserves IFR for all $n \geq n_d$.*

Due to the simple forms of $N_S(i,n,2)$ and $N_S(i,n,3)$, Hwang and Yao [120] showed that the results above can be used to give some conclusions of IFR for consecutive k-out-of-n system for $k=2$ and 3.

We now consider the case $k=2$. The following lemma is useful in the discussion.

LEMMA 4.3.12 *If $a_0, a_1, \cdots, a_m$ $(m \geq 0)$ satisfy $\sum_{i=0}^{j} a_i \geq 0$ for $j = 0, 1, \cdots, m$, then $\sum_{i=0}^{m} a_i p^i \geq 0$ for $0 \leq p \leq 1$.*

THEOREM 4.3.13 *$L(n,2)$ and $C(n,2)$ do not preserve IFR except for $L(2,2)$, $L(4,2)$, $L(6,2)$, $L(8,2)$, $C(2,2)$, $C(3,2)$, $C(4,2)$, $C(5,2)$, $C(7,2)$, $C(9,2)$ and $C(11,2)$.*

Proof. We only prove the theorem for $L(n,2)$ as the proof for $C(n,2)$ is similar. We consider two subcases.

Case 1. n is even, say $n=2q$. In this case,

$$\begin{aligned} \triangle &= N_L(q-1,2q,2) - qN_L(q,2q,2) \\ &= \binom{q+2}{q-1} - q\binom{q+2}{q-1} = \frac{q(q+1)(q-4)}{6} > 0 \quad \text{when } q \geq 5. \end{aligned}$$

According to Theorem 4.3.3, $L(n,2)$ does not preserve IFR for even $n \geq 10$.

Case 2. n is odd, say $n = 2q+1$. In this case

$$\begin{aligned}\Delta &= N_L(q, 2q+1, 2) - (q+1)N_L(q+1, 2q+1, 2)\\ &= \binom{q+2}{q} - (q+1)\binom{q+2}{q+1} = \frac{q(q+1)}{2} > 0 \text{ when } q \geq 1.\end{aligned}$$

Again, by Theorem 4.3.3, $L(n,2)$ does not preserve IFR for odd $n \geq 3$.

The exceptional cases can be verified by Theorem 4.3.2. We use $G_S(n,k)$ and $G'_S(n,k)$ to denote $G_S(p)$ and $G'_S(p)$ for $S = S(n,k)$. As we are only concerned with the sign of $G'_S(n,k)$, we use the symbol "$\propto$" to indicate the dropping of some positive constant factor in $G_S(n,k)$ and the dropping of some positive denominator in $G'_S(n,k)$. In every case, Lemma 4.3.12 is used to claim nonnegativity. We only do the case $L(4,2)$ as an example.

$$\begin{aligned}G_L(4,2) &= \frac{6p(1-p)+6p^2}{(1-p^2)+4p(1-p)+3p^2} \propto \frac{p}{1+2p},\\ G'_L(4,2) &\propto 1+2p-2p = 1 > 0.\end{aligned}$$

The proof is complete. ∎

Similarly, they also solve the $k = 3$ case.

THEOREM 4.3.14 *$L(n,3)$ and $C(n,3)$ do not preserve IFR except for $L(3,3)$, $L(4,3)$, $L(6,3)$, $L(9,3)$, $C(3,3)$, $C(4,3)$, $C(5,3)$, $C(7,3)$, $C(8,3)$, $C(10,3)$ and $C(13,3)$.*

As Derman, Lieberman and Ross [68] pointed out that for the consecutive k-out-of-n system not to preserve IFR, the intuitive meaning is there are possible states of each system where the potential for additional life of the system is greater in a state where more components have already failed than in a state where fewer components have failed.

For instance, for $L(n=3, k=2)$, where component lifetimes are exponentially distributed, the state $\bar{1}2\bar{3}$ (where i ($\bar{i}$) denotes component i in working (not working)) has a stochastically longer future life than the state $1\bar{2}3$; for $C(n=6, k=3)$, with component lives exponential, a similar statement holds for the state $1\bar{2}\bar{3}45\bar{6}$ and $\bar{1}2\bar{3}45\bar{6}$.

4.4 IFR PROPERTY FOR $k \geq 4$

Cui, Hawkes and Jalali [65] studied the IFR property for $k \geq 4$. They employed the formula for $N_L(d,n,k)$ in Theorem 2.3.4 to settle the following theorem which was Hwang and Yao's [120] conjecture.

THEOREM 4.4.1 *For every fixed k, there exists an n_k such that for every $n \geq n_k$ any consecutive k-out-of-n system, linear or circular, does not preserve IFR.*

Proof. First we consider $L(n,k)$ with $k \geq 3$. Let $n = kq + r$, where $0 \leq r < k$. By Theorem 2.3.4,

$$N_L(n-q-1,n,k) = \sum \binom{q+2}{t_1} \prod_{s=2}^{k-1} \binom{t_{s-1}}{t_s} \tag{4.4.1}$$

where the summation runs over $q+2 \geq t_1 \geq t_2 \geq \ldots \geq t_{k-1} \geq t_k = 0$ and

$$\sum_{s=1}^{k-1} t_s = n-q-1 = (k-1)q+r-1. \tag{4.4.2}$$

Then

$$\begin{aligned} t_{k-1} &= n-q-1-\sum_{s=1}^{k-2} t_s \\ &\geq n-q-1-(k-2)(q+2) = q+r-2k+3 \end{aligned} \tag{4.4.3}$$

with equality when $t_{k-2} = q+2$. For any fixed k and r, the right-hand side of this inequality is positive for all large enough q. The number of terms in summation (4.4.1) is then the number of vectors $(t_1, t_2, \ldots, t_{k-1})$ satisfying Eq. (4.4.2) with $q+2 \geq t_1 \geq \cdots \geq t_{k-1} \geq q+r+3-2k$. Now let $v_s = t_s - q$, and note that this is the same as the number of vectors $(v_1, v_2, \ldots, v_{k-1})$ satisfying $2 \geq v_1 \geq \cdots \geq v_{k-1} \geq r+3-2k$ and $\sum_{s=1}^{k-1} v_s = r-1$, which clearly does not depend on q.

Now, any individual term in summation (4.4.1) can be written as

$$\frac{(q+2)!}{t_{k-1}!} \frac{1}{(q+2-t_1)!} \prod_{s=2}^{k-1} \frac{1}{(t_{s-1}-t_s)!} \tag{4.4.4}$$

and each of $q+2-t_1$ and $t_{s-1}-t_s$ is at most $q+2-(q+r-2k+3) = 2k-r-1$, independently of q. Therefore, for fixed r and k, these can be taken as constants, and the form of the expression in Eq. (4.4.4) is governed by its first term, $(q+2)!/t_{k-1}!$, which is a polynomial in q of degree $q+2-t_{k-1}$. This is a maximum when t_{k-1} is smallest, as given by Eq. (4.4.3); i.e., we have a polynomial whose maximum degree is $q+2-(q+r-2k+3) = 2k-r-1$.

Thus, for q sufficiently large and fixed r and k ($k \geq 3$), we have shown that $N_L(n-q-1,n,k)$ is a polynomial in q of degree $2k-r-1$ with positive coefficients. In a similar way, we can show that $N_L(n-q,n,k)$ is a polynomial in q of degree $k-r-1$. Then, $N_L(n-q-1,n,k)-(n-q)N_L(n-q,n,k)$ is also a polynomial in q of degree $2k-r-1$ with a positive leading coefficient.

It follows that for any fixed k and for each r there is a $q(k,r)$ such that $\Delta = N_L(n-q-1,n,k)-(n-q)N_L(n-q,n,k) > 0$ for all $q \geq q(k,r)$. As this is true for each $0 \leq r \leq k-1$, there exists an n_k such that $\Delta > 0$ for all $n \geq n_k$. The theorem for $L(n,k)$ then follows from Theorem 4.3.3.

For the $C(n,k)$ case, using the same method and

$$N_C(i,n,k) = \frac{n}{n-i} N_L(i,n-1,k)$$

and remembering that now $n = kq - r$, we find for fixed k and r that $N_C(n-q-1,n,k)$ is a polynomial in q of degree $k+r$ with positive coefficients while $N_C(n-q,n,k)$ is a polynomial of degree r. Then, $N_C(n-q-1,n,k)-(n-q)N_L(n-q,n,k)$ is also a polynomial in q of degree $k+r$ with a positive leading coefficient. The same argument as above also proves the theorem for $C(n,k)$. ■

THEOREM 4.4.2 *$L(n,4)$ and $C(n,4)$ do not preserve IFR except for $L(4,4)$, $L(5,4)$, $L(8,4)$, $L(12,4)$, $C(4,4)$, $C(5,4)$, $C(6,4)$, $C(7,4)$, $C(9,4)$, $C(10,4)$, $C(13,4)$ and $C(17,4)$.*

Proof. Using Eq. (4.4.1) it is straightforward to write the expressions for particular $N_L(i,n,k)$. We consider four cases for $L(n,4)$.

Case 1. $n = 4q$. We get after some simplification that

$$N_L(n-q-1,n,4) = 6\binom{q+2}{3}+16\binom{q+2}{4}+15\binom{q+2}{5}+6\binom{q+2}{6}+\binom{q+2}{7},$$

$$N_L(n-q,n,4) = \binom{q+1}{1}+2\binom{q+1}{2}+\binom{q+1}{3},$$

$$\Delta = N_L(n-q-1,n,4)-(n-q)N_L(n-q,n,4) > 0 \text{ when } q \geq 4.$$

Case 2. $n = 4q+1$. In this case.

$$N_L(n-q-1,n,4) = \binom{q+2}{2}+7\binom{q+2}{3}+10\binom{q+2}{4}+5\binom{q+2}{5}+\binom{q+2}{6},$$

$$N_L(n-q,n,4) = \binom{q+1}{1}+\binom{q+1}{2},$$

$$\Delta = N_L(n-q-1,n,4)-(n-q)N_L(n-q,n,4) > 0 \text{ when } q \geq 2.$$

Case 3. $n = 4q+2$. Now,

$$N_L(n-q-1, n, 4) = 2\binom{q+2}{2} + 6\binom{q+2}{3} + 4\binom{q+2}{4} + \binom{q+2}{5},$$

$$N_L(n-q, n, 4) = q+1,$$

$$\Delta = N_L(n-q-1, n, 4) - (n-q)N_L(n-q, n, 4) > 0 \text{ when } q \geq 1.$$

Case 4. $n = 4q+3$. We have

$$N_L(n-q-1, n, 4) = 3\binom{q+2}{2} + 3\binom{q+2}{3} + \binom{q+2}{4},$$

$$N_L(n-q, n, 4) = 1,$$

$$\Delta = N_L(n-q-1, n, 4) - (n-q)N_L(n-q, n, 4) > 0 \text{ when } q \geq 1.$$

Theorem 4.3.3 implies that all the cases where $\Delta > 0$ do not preserve IFR. Note that all the cases dealt with under Case 4 are already known not to preserve IFR by Theorem 4.3.6. We now verify the exceptional cases. We know that $L(4,4)$, $L(5,4)$, and $L(8,4)$ do preserve IFR from Theorem 4.3.7. This leaves only $L(12,4)$ in doubt. For this case, we have

$$G_L(p) = \frac{36p^3 + 32p^4 + 28p^5 + 24p^6 - 60p^7}{1 + 3p + 6p^2 + 10p^3 + 6p^4 + 2p^5 - 2p^6 - 6p^7}.$$

Using the symbol $\propto$ to indicate dropping the positive denominator in the derivative $G'_L(p)$, we have

$$\begin{aligned} G'_L &\propto 108p^2 + 344p^3 + 644p^4 + 864p^5 + 548p^6 - 88p^7 - 760p^8 \\ &\quad -1120p^9 - 400p^{10} + 96p^{11} + 264p^{12} \\ &= (1-p)(108p^2 + 452p^3 + 1096p^4 + 1960p^5 + 2508p^6 + 2420p^7 \\ &\quad +1660p^8 + 540p^9 - 140p^{10} + 236p^{11}) + 500p^{12} \geq 0. \end{aligned}$$

Thus, $G_L(p)$ is an increasing function of p for $0 \leq p < 1$ and Theorem 4.3.2 implies that $L(12,4)$ preserves IFR. This completes the proof for the linear case.

The proof for the circular case is similar and thus omitted. ∎

The results of this theorem show that Theorem 4.4.1 holds for $k = 4$ with IFR being not preserved for all $n \geq n_4 = 13$ in $L(n,4)$ and for all $n \geq n_4 = 18$ in $C(n,4)$. Note also that the preceding expressions in Cases 1 to 4 are indeed polynomials in q whose degrees are those predicted by the proof of Theorem 4.4.1.

THEOREM 4.4.3 *$L(n,5)$ and $C(n,5)$ do not preserve IFR except for $L(5,5)$, $L(6,5)$, $L(7,5)$, $L(10,5)$, $L(11,5)$, $L(15,5)$, $C(5,5)$, $C(6,5)$, $C(7,5)$, $C(8,5)$, $C(9,5)$, $C(11,5)$, $C(12,5)$, $C(13,5)$, $C(16,5)$, $C(17,5)$ and $C(21,5)$.*

The proof is similar to that for Theorem 4.4.2, so the details are omitted. We see that $L(n,5)$ does not preserve IFR for all $n \geq n_5 = 16$ while $C(n,5)$ does not preserve IFR for all $n \geq n_5 = 22$. Note also that IFR is preserved for $L(7,5)$ and $C(13,5)$, for each of which $\Delta = 0$; the other mentioned cases all have negative Δ's.

Chapter 5

ASYMPTOTIC ANALYSIS

In this chapter we study the reliabilities of consecutive systems when $n \to \infty$. If every component has a constant failure probability independent of n, then the reliability of the system tends to zero as $n \to \infty$ which is of no interest. Therefore, in the following we assume that the component probabilities depend on n.

5.1 ELEMENTARY METHOD

Fu [77] used elementary probabilistic analysis to prove the following result which was conjectured by Chao and Lin [52].

THEOREM 5.1.1 *For any integer $k \geq 1$, if the component reliability $p_n = 1 - \lambda n^{-1/k}$ ($\lambda > 0$) holds for IID model, then the consecutive k-out-of-n system reliability*

$$R_L(n) \approx \exp\{-nq_n^k\} \text{ as } n \to \infty, \text{ i.e., } \lim_{n\to\infty} R_L(n) = \exp\{-\lambda^k\}.$$

Proof. It follows from Theorem 2.4.9 that

$$(1 - q_n^k)^{n-k+1} \leq R_L(n) \leq (1 - p_n q_n^k)^{n-k+1}.$$

Note that $q_n = \lambda / n^{1/k}$. So,

$$\lim_{n\to\infty} (1 - q_n^k)^{n-k+1} = \lim_{n\to\infty} \left(1 - \frac{\lambda^k}{n}\right)^{n-k+1} = \exp\{-\lambda^k\} \quad \text{and}$$

$$\lim_{n\to\infty} (1 - p_n q_n^k)^{n-k+1} = \lim_{n\to\infty} \left(1 - \frac{\lambda^k}{n} + \frac{\lambda^{k+1}}{n^{1+1/k}}\right)^{n-k+1} = \exp\{-\lambda^k\}.$$

By the sandwich theorem for limit, $\lim_{n\to\infty} R_L(n) = \exp\{-\lambda^k\}$. ■

For the consecutive k-out-of-n system, Chao and Fu [50] embedded the system in a finite Markov chain $\{X_0, X_1, \ldots, X_n\}$ with state space $S = \{1, 2, \ldots, k+1\}$ and transition probability matrix

$$\Lambda_n(t) = \begin{pmatrix} p_t & q_t & & & \\ p_t & & q_t & & \\ \vdots & & & \ddots & \\ p_t & & & & q_t \\ 0 & & & & 1 \end{pmatrix}_{(k+1)\times(k+1)},$$

where the (i,j)th cell of $\Lambda_n(t)$ is $p_{ij}(t,n) = \Pr\{X_t = j | X_{t-1} = i\}$. Let $\pi_0 = (\pi_1, \pi_2, \ldots, \pi_{k+1})$ be the initial probability vector, which is $(1, 0, \ldots, 0)$ usually. Then the system reliability is

$$R_L(n) = \pi_0 \prod_{t=1}^{n} \Lambda_n(t) U_0,$$

where U_0 is the $(k+1) \times 1$ column vector $(1, \ldots, 1, 0)^{\mathrm{T}}$.

Let $b(t,n) = \Pr\{X_t = k+1 | X_{t-1} \le k\}$, for $1 \le t \le n$, be the conditional probability that given the system is working at time $t-1$ and the system breaks down at time t; and $a_j(t,n) = \Pr\{X_t = j\}$, for $1 \le t \le n$, be the probability that the system is in state j at time t. Then

$$a(t,n) = (a_1(t,n), a_2(t,n), \ldots, a_{k+1}(t,n)) = \pi_0 \prod_{j=1}^{t} \Lambda_n(j).$$

Also,

$$b(t,n) = \frac{a(t-1,n)W(t,n)}{a(t-1,n)U_0},$$

where $W(t,n)$ is the $(k+1) \times 1$ vector $(p_{1(k+1)}(t,n), \ldots, p_{k(k+1)}(t,n), 0)^{\mathrm{T}}$. Note that $p_{1j}(t,n) = 0$ if $j - 1 \ge 2$.

LEMMA 5.1.2 *For a linear system,*

$$R_L(n) = \prod_{i=1}^{n} (1 - b(i,n)).$$

Proof. Note that

$$\Lambda_n(t)U_0 = (1 - p_{1(k+1)}(t), \ldots, 1 - p_{k(k+1)}(t), 0)^{\mathrm{T}} = U_0 - W(t,n).$$

Further, by the Chapman-Kolmogorov equation,

$$\begin{aligned}
R_L(n) &= \pi_0 \prod_{t=1}^{n} \Lambda_n(t) U_0 \\
&= a(n-1,n)\Lambda_n(n)U_0 \\
&= a(n-1,n)(U_0 - W(n,n)) \\
&= \left(1 - \frac{a(n-1,n)W(t,n)}{a(n-1,n)U_0}\right) a(n-1,n)U_0 \\
&= (1-b(n,n))a(n-1,n)U_0 \\
&= (1-b(n,n))(1-b(n-1,n))a(n-2,n)U_0 \\
&\vdots \\
&= \prod_{i=1}^{n}(1-b(i,n))a(0,n)U_0 \\
&= \prod_{i=1}^{n}(1-b(i,n)),
\end{aligned}$$

where $a(0,n) = \pi_0$ and $\pi_0 U_0 = 1$. ■

Denote

$$\lambda_n = \sum_{j=1}^{\infty} \frac{1}{j} \sum_{t=1}^{n} b^j(t,n).$$

Condition A. Given π_0, there exists $\lambda > 0$ such that

$$\lim_{n\to\infty} \lambda_n = \lambda.$$

THEOREM 5.1.3 *If condition A is satisfied and $\pi_0 U_0 = 1$, then*

$$\lim_{n\to\infty} R_L(n,k) = \exp\{-\lambda\}.$$

Proof. Since $\pi_0 U_0 = 1$, it follows from Lemma 5.1.2 that

$$R_L(n) = \prod_{i=1}^{n}(1-b(i,n)).$$

Taking log on both sides, it follows from Taylor's expansion of $\log(1-x)$ that

$$\log R_L(n) = \sum_{i=1}^{n} \log(1-b(i,n)) = -\sum_{j=1}^{\infty} \frac{1}{j} \sum_{t=1}^{n} b^j(t,n) = -\lambda_n.$$

The theorem follows immediately from Condition A and the above equalities. ■

5.2 GENERATING FUNCTION METHOD

Using the generating function, Papastavridis [187] gave the following theorem for a consecutive k-out-of-n system.

Let T be a random variable which is the time of a component's failure and let $q(t) = \Pr\{T \leq t\}$ for $t \geq 0$ be component's failure distribution. Let T_n be the system's time to failure.

THEOREM 5.2.1 *If* $q(t) = (\lambda t)^\alpha + o(t^\alpha)$, *where* α *and* λ *are positive real constants, then*

$$\lim_{n\to\infty} \Pr\{n^{1/k\alpha}T_n \leq t\} = 1 - \exp\{-(\lambda t)^{k\alpha}\}$$

for all $t \geq 0$.

Proof. Let $t_n = tn^{-1/k\alpha}$ and let $x(t)$ be the unique positive root of the polynomial $1 - pz(1 + qz + \ldots + q^{k-1}z^{k-1})$.

Note that the system reliability for $p(t_n)$ is $R_L(n) = \Pr\{T_n > t_n\}$. In order to prove the theorem, it is the same to prove that

$$\lim_{n\to\infty} R_L(n) = \exp\{-(\lambda t)^{k\alpha}\}. \tag{5.2.1}$$

It follows from Feller [75], page 325, that

$$R_L(n) \sim \frac{1 - q(t_n)x(t_n)}{(k + 1 - kx(t_n))p(t_n)} x(t_n)^{-(n+1)}. \tag{5.2.2}$$

But as $t_n \to 0$, it follows that $q(t_n) \to 0$ and $p(t_n) \to 1$, as $n \to \infty$. From Feller [75], page 326, (7.19), we know that

$$x(t) = 1 + p(t)q(t)^k + (k+1)(p(t)q(t)^k)^2 + \ldots$$

if $(k+1)p(t) > 1$. Since $q(t) = (\lambda t)^\alpha + o(t^\alpha)$, we have

$$x(t) = 1 + (\lambda t)^{k\alpha} + o(t^{k\alpha}) \quad \text{and} \quad \log x(t) = (\lambda t)^{k\alpha} + o(t^{k\alpha}).$$

By substitution, we get

$$\log x(t_n) = (\lambda t)^{k\alpha}/n + o(1/n) \to 0$$

and

$$-\frac{\log x(t_n)}{1/(n+1)} = -\frac{(\lambda t)^{k\alpha}/n + o(1/n)}{1/(n+1)} \to -(\lambda t)^{k\alpha}.$$

Taking these remarks into consideration, we get (5.2.1) from (5.2.2). This completes the proof of the theorem. ∎

Let T_n^c be the lifetime of a circular consecutive k-out-of-n:F system. Obviously, we have

$$\Pr\{T_n \le t\} \le \Pr\{T_n^c \le t\} \le \Pr\{T_n \le t\} + \sum_{j=n-k+2}^{n} \prod_{i=j}^{j+k-1} q_i(t). \quad (5.2.3)$$

Hence, if $\sum_{j=n-k+2}^{n} \prod_{i=j}^{j+k-1} q_i(t) \to 0$ under some conditions, then a limit theorem for lines also holds for cycle; see the discussion after Theorem 5.3.2.

5.3 POISSON CONVERGENCE METHOD

When the component reliabilities are not equal, Chryssaphinou and Papastavridis [63] gave the following results by using the theorem of Barbour and Eagleson [13] who studied Poisson convergence via the Stein-Chen method.

Let F_i be the failure distribution of the ith component. Define

$$\begin{aligned} P_j(t) &= \prod_{i=j}^{j+k-1} F_i(t) \quad (1 \le j \le n-k+1), \\ p(t) &= \max_{1 \le i \le n} F_i(t), \\ \lambda(t) &= \sum_{j=1}^{n-k+1} P_j(t). \end{aligned}$$

Lemma 5.3.1

$$|\Pr\{T \le t\} - (1 - e^{-\lambda(t)})| \le (2k-1)p(t)^k + (2k-2)p(t).$$

Proof. Consider the random variable

$$X_j = \begin{cases} 1 & \text{if components } j, j+1, \ldots, j+k-1 \text{ fail}, \\ 0 & \text{otherwise} \end{cases}$$

for $1 \le j \le n-k+1$ and the random variable $X = \sum_{j=1}^{n-k+1} X_j$. It is clear that the system fails if and only if $X > 0$. From the definitions of X_j and X, we have $E(X_j) = P_j(t) \quad (1 \le j \le n-k+1)$ and $E(X) = \sum_{j=1}^{n-k+1} P_j(t) = \lambda(t)$. According to Theorem 2 of Barbour and Eagleson [13], we have

$$|\Pr\{T \le t\} - (1 - e^{-\lambda(t)})|$$

$$\leq \min\left(1, 1/E(X)\right) \sum_{j=1}^{n-k+1} \left(P_j(t)^2 + \sum_{\substack{i=j-k+1 \\ i \neq j}}^{j+k-1} \left(P_j(t)P_i(t) + E(X_jX_i) \right) \right)$$

$$\leq \left(1/E(X)\right) \left(\left(\sum_{j=1}^{n-k+1} P_j(t)p(t)^k \right) + \left(\sum_{j=1}^{n-k+1} P_j(t) \right) (2k-2)p(t)^k \right.$$

$$\left. + \left(\sum_{j=1}^{n-k+1} E(X_j) \right) (2k-2)p(t) \right)$$

$$= (2k-1)p(t)^k + (2k-2)p(t).$$

Note that the second inequality follows from

$$\begin{aligned} E(X_jX_i) &= \Pr\{X_j = X_i = 1\} \\ &= \Pr\{X_j = 1\}F_i(t)\dots F_{j-1}(t) \\ &\leq \Pr\{X_j = 1\}p(t) = E(X_j)p(t). \end{aligned}$$

This completes the proof of the lemma. ■

THEOREM 5.3.2 *Suppose the following three conditions holds.*

(i) *There are positive numbers λ_i, α_i and functions $\phi_i(t)$ such that $F_i(t) = (\lambda_i t)^{\alpha_i} + t^{\alpha_i}\phi_i(t)$ $(i = 1, 2, \dots)$ for $0 \leq t < \delta$, where δ is a positive number.*

(ii) $\lim_{t \to 0} \phi_i(t) = 0$ *uniformly on i.*

(iii) $\lim_{i \to \infty} \lambda_i = \lambda$.

Let $\alpha = \inf \alpha_i$.

(a) *If $\alpha = \alpha_i$ for $i = 1, 2, \dots,$ then*

$$\lim_{n \to \infty} \Pr\{n^{1/k\alpha} T_n \leq t\} = 1 - \exp\{-(\lambda t)^{k\alpha}\}$$

(b) *If $\alpha > 0$ and for every $i = 1, 2, \dots,$ there is a j with $i \leq j \leq i+k-1$ so that $\alpha_j > \alpha$, then*

$$\lim_{n \to \infty} \Pr\{n^{1/k\alpha} T_n \leq t\} = 0.$$

Proof. According to conditions (i), (ii) and (iii), we have

$$p(tn^{-1/k\alpha}) = \max_{1 \leq i \leq n} \left(t^{\alpha_i} n^{-\alpha_i/k\alpha} (\lambda_i^\alpha + \phi_i(tn^{-1/k\alpha})) \right) \to 0 \quad \text{as } n \to \infty.$$

On the other hand,

$$\lambda(tn^{-1/k\alpha}) = \sum_{j=1}^{n-k+1} \prod_{i=j}^{j+k-1} \left(t^{\alpha_i} n^{-\alpha_i/k\alpha} (\lambda_i^\alpha + \phi_i(tn^{-1/k\alpha})) \right)$$

$$\to \begin{cases} (\lambda t)^{k\alpha} & \text{as } n \to \infty \text{ for (a);} \\ 0 & \text{as } n \to \infty \text{ for (b).} \end{cases}$$

The theorem then follows from Lemma 5.3.1. ■

For the circular case, the theorem also holds due to Eq. (5.2.3). Theorem 5.3.2 is applicable to the following two examples.

Example 1. The life length distribution of the ith component is a Weibull distribution, i.e.,

$$F_i(t) = 1 - \exp\{-(\lambda_i t)^\alpha\} = (\lambda_i t)^\alpha + t^\alpha o(1)$$

with $\lambda_i \to \lambda$ as $i \to \infty$.

Example 2. The life length distribution consists of two mixed kinds of distributions Weibull and Gamma distributions, i.e.,

$$G_{2i-1}(t) = 1 - \exp\{-(\lambda_i t)^\alpha\} = (\lambda_i t)^\alpha + t^\alpha o(1)$$

and

$$G_{2i}(t) = \int_0^t \frac{\mu_i^\alpha}{\Gamma(\alpha)} s^{\alpha-1} e^{-\mu_i s} ds = \frac{\mu_i^\alpha}{\Gamma(\alpha+1)} t^\alpha + t^\alpha o(1)$$

with $\lambda_i \to \lambda$ and $\mu_i \to \lambda(\Gamma(\alpha+1))^{1/\alpha}$ as $i \to \infty$.

5.4 THE DEPENDENCE MODELS

Fu [79] first considered the case that the component states are dependent. In particular, he assumed that the failure probability of a component in a consecutived-k-out-of-n line depends only on the number of consecutive failed components immediately preceding it. Thus if component i is preceded by exactly m consecutive failed components for $0 \le m \le k-1$, then $q_i = q(m)$. Note that the definition of q_i is immaterial if $m \ge k$ since the system is failed before reaching component i

Let $R_L(m; i)$ denote the reliability of the system up to component i given that component i is preceded by exactly m consecutive failed components. Fu [79] gave

THEOREM 5.4.1 *$R_L(i) = \sum_{m=0}^{k-1} R_L(m; i)$ for $i = 1, 2, \ldots, n$, where $R_L(m; i)$ is computed by the recursive equations*

$R_L(0, 0) = 1$, $R_L(0, i) = \sum_{m=0}^{k-1} p(m) R(m; i-1)$ for $1 \le i \le k-1$.

$R_L(m; i) = 0$ for $i \le m \le k-1$,

$R_L(m;i) = \prod_{j=0}^{m-1} q(j) \sum_{t=1}^{k-1} p(t)R(t;i-m-1)$ *for* $1 \le m \le k-1, m \le i \le n$.

Proof. Straightforward. ■

Fu and Hu [81] showed that if the component failure probabilities $q(m)$, $m = 0, 1, \ldots, k-1$, are functions of n and tend to zero with the rates,

$$q(m) = \lambda_m n^{-1/k}, \lambda_m > 0, \text{ for } m = 0, 1, \ldots, k-1,$$

then

$$R_L(n) \approx \exp\left(-n \prod_{m=0}^{k-1} q(m)\right) \text{ as } n \to \infty$$

and

$$R_L(\infty) \equiv \lim_{n\to\infty} R_L(n) = \exp\left(-\prod_{m=0}^{k-1} \lambda_m\right).$$

Papastavridis [187] considered the 1-step Markov process, i.e., the probability of component i failed depends on the state of component $i-1$ but not on the states of other components. Let X_i be zero or one as component i fails or works $i = 1, 2, \ldots, n$. Then

$$p_{i,0} = \Pr\{X_i = 1 | X_{i-1} = 0\} \quad \text{and} \quad p_{i,1} = \Pr\{X_i = 1 | X_{i-1} = 1\}$$

for $2 \le i \le n$ and

$$p_{1,0} = p_{1,1} = \Pr\{X_1 = 1\}.$$

He proved

THEOREM 5.4.2

$$R(n) = \begin{cases} 1, & \text{if } n < k; \\ R(n-1) + \frac{q_{n,0}q_{n-k+1,1}}{q_{n,0}q_{n-1,0}}(p_{n-k,1} - p_{n-k,0}) & \\ (R(n-1) - R(n-2)) - (q_{n,0}q_{n-1,0}\cdots q_{n-k+2,0}) & \\ q_{n-k+1,1}p_{n-k,0}R(n-k-1), & \text{if } n \ge k, \end{cases}$$

where $q_{i,0} = 1 - p_{i,0}$ *and* $q_{i,1} = 1 - p_{i,1}$ *for* $1 \le i \le n$.

Proof. For convenience, define

$R_1(n) = \Pr\{\text{the system works and component } n \text{ works}\}$

and

$R_0(n) = \Pr\{\text{the system works but component } n \text{ fails}\}$.

Note that

$$R(n) = R_1(n) + R_0(n).$$

Since $X_1, X_2, \ldots, X_n$ form a Markov chain and

$$\Pr\{X_i = j_i | X_1 = j_1, \ldots, X_{i-1} = j_{i-1}\} = \Pr\{X_i = j_i | X_{i-1} = j_{i-1}\},$$

we have

$$R_1(n) = p_{n,0}R(n-1) + (p_{n,1} - p_{n,0})R_1(n-1), \tag{5.4.1}$$

$$R(n) = R(n-1) - (q_{n,0}q_{n-1,0}\cdots q_{n-k+2,0})q_{n-k+1,1}R_1(n-k). \tag{5.4.2}$$

Then

$$\begin{aligned} R(n) &= R(n-1) - (q_{n,0}q_{n-1,0}\cdots q_{n-k+2,0})q_{n-k+1,1}R_1(n-k) \\ &= R(n-1) - (q_{n,0}q_{n-1,0}\cdots q_{n-k+2,0})q_{n-k+1,1}(p_{n-k,0} \\ &\quad R(n-k-1) + (p_{n-k,1} - p_{n-k,0})R_1(n-k-1)) \quad \text{by (5.4.1)} \\ &= R(n-1) - (q_{n,0}q_{n-1,0}\cdots q_{n-k+2,0})q_{n-k+1,1}p_{n-k,0}R(n-k-1) \\ &\quad +(q_{n,0}q_{n-1,0}\cdots q_{n-k+2,0})q_{n-k+1,1}(p_{n-k,1} - p_{n-k,0}) \\ &\quad \frac{R(n-1) - R(n-2)}{(q_{n-1,0}q_{n-2,0}\cdots q_{n-k+1,0})q_{n-k,1}} \quad \text{by (5.4.2)} \\ &= R(n-1) + \frac{q_{n,0}q_{n-k+1,1}}{q_{n-k+1,0}q_{n-k,1}}(p_{n-k,1} - p_{n-k,0})(R(n-1) - \\ &\quad R(n-2)) - (q_{n,0}q_{n-1,0}\cdots q_{n-k+2,0})q_{n-k+1,1}p_{n-k,0}R(n-k-1). \end{aligned}$$

■

Ge and Wang [83] presented a direct method for the system reliability when $p_{i,1} = p_1, p_{i,0} = p_0, p_{0,0} = p$.

LEMMA 5.4.3 *Let $M(i,m,k)$ be the number of ways of placing i identical balls in m different urns, with each urn containing at least one ball and at most k balls, then*

$$M(i,m,k) = \sum_{j=0}^{m^*}(-1)^j \binom{m}{j}\binom{i-kj-1}{m-1}$$

for $i \geq m > 0$, where $m^* = \min(m, \lfloor (i-m)/k \rfloor)$.

Proof. Since $M(i,m,k)$ is the coefficient of t^i in $(t + t^2 + \ldots + t^k)^m$, we have

$$\begin{aligned} (t + t^2 + \ldots + t^k)^m &= \left(\frac{t - t^{k+1}}{1-t}\right)^m \\ &= t^m(1-t^k)^m(1-t)^{-m} \end{aligned}$$

$$= t^m \sum_{j=0}^{m} (-1)^j t^{kj} \binom{m}{j} \sum_{p=0}^{\infty} \binom{m+p-1}{p} t^p$$

$$= \sum_{i=0}^{\infty}\sum_{j=0}^{\infty} (-1)^j \binom{m}{j} \binom{i-kj-1}{m-1} t^i.$$

This proves the lemma. ■

Let $M(i,m)$ be the number of ways to place i identical balls in m different urns, with each urn containing at least one ball. Then

$$M(i,m) = M(i,m,i-m+1) = \binom{i-1}{m-1}.$$

Ge and Wang [83] then proved

THEOREM 5.4.4 *$R(n) = pp_1^{n-1} + R_1 + R_2 + R_3 + R_4$, where*

$$R_1 = \sum_{i=1}^{n-1} \sum_{m=1}^{\min(i,n-i)} M(i,m,k-1)M(n-i,m)p_0^m p_1^{n-i-m} q q_0^{i-m} q_1^{m-1},$$

$$R_2 = \sum_{i=1}^{n-1} \sum_{m=1}^{\min(i,n-i)} M(i,m,k-1)M(n-i,m)pp_0^{m-1} p_1^{n-i-m} q_0^{i-m} q_1^{m},$$

$$R_3 = \sum_{i=1}^{n-2} \sum_{m=1}^{\min(i,n-i-1)} M(i,m,k-1)M(n-i,m+1)pp_0^{m} p_1^{n-i-m-1} q_0^{i-m} q_1^{m},$$

$$R_4 = \sum_{i=2}^{n-1} \sum_{m=2}^{\min(i,n-i+1)} M(i,m,k-1)M(n-i,m-1)p_0^{m-1} p_1^{n-i-m+1} q q_0^{i-m} q_1^{m}.$$

Proof. Since the states of the system consist of working and failed strings of components. There are four possible cases for the system states, besides the case that all components work:

(1) The number of failed strings equals the number of working strings and the system states begin with a failed string.

(2) The number of failed strings equals the number of working strings and the system states begin with a working string.

(3) The number of working strings equals the number of failed strings plus one.

(4) The number of working strings equals the number of failed strings minus one.

Let $R_i, i = 1,2,3,4$, denote the reliability of the above cases. Then it can be verified that R_1, R_2, R_3, R_4 are as given in the theorem. ■

Under the same assumption of Papastavridis [187], Boland, Proshan and Tong [26] presented another form of dependence when $p_{i,0}$ and $p_{i,1}$ are constants for $i = 1, 2, \ldots, n$. They showed that

$$p_{i,0} = (1-r)p \quad \text{and} \quad p_{i,1} = r + (1-r)p;$$

$$\text{i.e., } r = p_{i,1} - p_{i,0} \quad \text{and} \quad p = p_{i,0}/(q_{i,1} + p_{i,0}).$$

Namely, the dependence can be described by using two independent parameters r and p.

Let $U_1, U_2, \ldots, U_n$ be independent Bernoulli random variables with

$$\Pr\{U_i = 1\} = p \quad \text{and} \quad \Pr\{U_i = 0\} = 1 - p \quad (i = 1, 2, \ldots, n)$$

and $X_1, X_2, \ldots, X_n$ be Bernoulli random variables defined by $X_1 = U_1$ and

$$X_i = \begin{cases} X_{i-1}, & \text{with probability } r; \\ U_i, & \text{with probability } 1 - r \end{cases}$$

for $i = 2, 3, \ldots, n$, where $r \in [0, 1]$ is fixed. It is clear that the random variables $X_1, X_2, \ldots, X_n$ are dependent (except for the case where $r = 0$). Also, the sequence $X_1, X_2, \ldots, X_n$ has the Markovian property that

$$\Pr\{X_j = x_j | X_{j_s} = x_{j_s}, \ldots, X_{j_1} = x_{j_1}\} = \Pr\{X_j = x_j | X_{j_s} = x_{j_s}\}$$

for any $j_1 < j_2 < \ldots < j_s < j$. They gave

THEOREM 5.4.5 (1) $\Pr\{X_i = 1\} = p$ *for* $i = 1, 2, \ldots, n$.

(2) $\text{corr}(X_i, X_j) = r^{|j-i|}$ *for all* i *and* j.

(3) $\Pr\{X_i = \varepsilon_i$ *for* $i = 1, 2, \ldots, n\} = \Pr\{X_i = \varepsilon_{n-i}$ *for* $i = 1, 2, \ldots, n\}$ *for any* 0-1 *sequence* $\varepsilon_1, \varepsilon_2, \ldots, \varepsilon_n$.

Proof. (1) We prove the assertion by induction on i. It is clear for $i = 1$. If $\Pr\{X_i = 1\} = p$, then

$$\Pr\{X_{i+1} = 1\} = r\Pr\{X_i = 1\} + (1-r)\Pr\{U_{i+1} = 1\} = rp + (1-r)p = p.$$

(2) Let $s = j - i$. By the Markovian property of the sequence, we have

$$\begin{aligned} &\Pr\{X_{i+s} = X_i = 1\} \\ &\quad = p\Pr\{X_{i+s} = 1 | X_i = 1\} \\ &\quad = p(r^s + r^{s-1}(1-r)p + r^{s-2}(1-r)p + \ldots + (1-r)p) \\ &\quad = p(r^s + p(1 - r^s)) = pqr^s + p^2. \end{aligned}$$

Hence $\text{corr}(X_{i+s}, X_i) = ((pqr^s + p^2) - p^2)/(pq) = r^s$.

(3) For any sequence $\varepsilon = (\varepsilon_1, \varepsilon_2, \ldots, \varepsilon_n)$, define G and F to be the number of runs of ones and zeros in the sequence, S be the number of ones. Note that for any i, we have $\Pr\{X_{i+1} = 0 | X_i = 1\} = (1-r)q$ and $\Pr\{X_{i+1} = 0 | X_i = 0\} = r + (1-r)q$. By individually inspecting sequences ε where $G = F, G = F+1$, and $G = F-1$, it may be established that

$$\begin{aligned}&\Pr\{X_1 = \varepsilon_1, X_2 = \varepsilon_2, \ldots, X_n = \varepsilon_n\}\\ &\quad = p^G q^F (1-r)^{G+F-1}(r+(1-r)p)^{S-G}(r+(1-r)q)^{n-S-F}.\end{aligned}$$

Now consider any sequence $\varepsilon = (\varepsilon_1, \varepsilon_2, \ldots, \varepsilon_n)$ of zeros and ones. It is clear that if $\varepsilon^t = (\varepsilon_n, \varepsilon_{n-1}, \ldots, \varepsilon_1)$, then $G(\varepsilon) = G(\varepsilon^t)$, $F(\varepsilon) = F(\varepsilon^t)$ and $S(\varepsilon) = S(\varepsilon^t)$. Therefore, it follows from the above probability expression that $\Pr\{\mathbf{X} = \varepsilon\} = \Pr\{\mathbf{X} = \varepsilon^t\}$. ■

THEOREM 5.4.6 *For $k = 2$, we have $R(n) =$*

$$\begin{aligned}\sum_{j=0}^{[(n+1)/2]} & p\left\{\binom{n-j-1}{j-1}((1-r)q)^j((1-r)p)^{j-1}(r+(1-r)p)^{n-2j} + \right.\\ &\left.\binom{n-j-1}{j}((1-r)p)^j((1-r)q)^j(r+(1-r)p)^{n-2j-1}\right\} +\\ & p\left\{\binom{n-j-1}{j-2}((1-r)q)^{j-1}((1-r)p)^{j-1}(r+(1-r)p)^{n-2j+1} + \right.\\ &\left.\binom{n-j-1}{j-1}((1-r)p)^j((1-r)q)^{j-1}(r+(1-r)p)^{n-2j}\right\}.\end{aligned}$$

Proof. This follows by considering separately the cases in which the system functions and there are $j = F$ failures and by considering the working and failed cases for the first and the last components. ■

When $r = 0$, the above theorem reduces to that of the case of independent components and

$$R(n) = \sum_{0}^{[(n+1)/2]} \binom{n-j+1}{j}(1-p)^j p^{n-j},$$

which is the expression obtained by Chiang and Niu [61].

THEOREM 5.4.7 *For $k \geq n/2$, the reliability of a consecutive k-out-of-n:F system is*

$$1 - \{\Pr(X_1 = \ldots = X_k = 0) + \sum_{i=1}^{n-k} \Pr(X_i = 1, X_{i+1} = \ldots = X_{i+k} = 0)\},$$

whenever the X_i are identically distributed Bernoulli variables with $E(X_i) = p$ (no matter what the dependence structure is).

Proof. For each $j = k, k+1, \dots, n$ and $s = 0, 1, \dots, n-j$, we define $A_{j,s}$ to be the event $\{X_s = 1, X_{s+1} = 0, \dots, X_{s+j} = 0, X_{s+j+1} = 1\}$. Here we let $X_0 \equiv X_{n+1} \equiv 1$. Since $k \geq n/2$, it follows that events $\{A_{j,s}\}$ are pairwise disjoint, and hence,

$$R(n) = 1 - \sum_{j=k}^{n} \sum_{s=0}^{n-j} \Pr\{A_{j,s}\} = 1 - \sum_{s=0}^{n-k} \sum_{j=k}^{n-s} \Pr\{A_{j,s}\}.$$

Since $\sum_{j=k}^{n-s} \Pr\{A_{j,s}\} = \Pr\{X_s = 1, X_{s+1} = 0, \dots, X_{s+k} = 0\}$, the theorem follows. ∎

COROLLARY 5.4.8 *For $k \geq n/2$,*

$$R(n) = 1 - q\Big(1 + (n-k)p(1-r)\Big)\Big(r + (1-r)p\Big)^{k-1}$$

Proof. The corollary follows from

$$\Pr\{X_1 = \ldots = X_k = 0\} = q\Big(r + (1-r)q\Big)^{k-1} \quad \text{and}$$

$$\Pr\{X_i = 1, X_{i+1} = \ldots = X_{i+k} = 0\} = p\Big((1-r)q\Big)\Big(r + (1-r)q\Big)^{k-1}.$$

∎

COROLLARY 5.4.9 *For $k \geq (n+1)/2, R(n)$ is a decreasing function of r. For $k = n/2$ (when n is even), $R(n)$ is a decreasing function of r on the interval $[0, \max\{0, 1 - 4/(n^2 p)\}]$ and an increasing function of r on the interval $[\max\{0, 1 - 4/(n^2 p)\}, 1]$.*

Proof. According to Corollary 5.4.8, we have for that $n \geq n/2$,

$$\frac{\partial}{\partial r} R(n) \leq 0 \Longleftrightarrow k - 1 + (n-k)\Big((k-1)p(1-r) - r - q(1-r)\Big) \geq 0.$$

Since $(k-1)p(1-r) - r - q(1-r) \geq -1 \Longleftrightarrow p(1-r)k$, which is true. Therefore, if $k \geq (n+1)/2$, then

$$k-1+(n-k)\Big((k-1)p(1-r)-r-q(1-r)\Big) \geq k-1-(n-k) = 2k-n-1 \geq 0.$$

Hence, the result for $k \geq (n+1)/2$. Now consider the case when $k = n/2$ (n is even). Then

$$k - 1 + (n-k)\Big((k-1)p(1-r) - r - q(1-r)\Big) = (n/2)^2 p(1-r) - 1 \geq 0$$

$$\iff r \le 1 - 4/(n^2 p).$$

Hence the result for $k = n/2$. ■

5.5 DISTRIBUTION FOR EXCHANGEABLE LIFETIMES

If the joint cdf of component lifetimes of the system is invariant under the permutation of the component indices, then we call the n components of the system *statistically exchangeable.* Shanthikumar [236] first studied the s-exchangeable problem to linear consecutive k-out-of-n:F system. Papastavridis [191] gave the solution to the same question for the circular system as follows. Let

$$p(j,t) = \Pr\{T_1 \le t, T_2 \le t, \ldots, T_j \le t, T_{j+1} \ge t, \ldots, T_n \ge t\}.$$

THEOREM.5.5.1 *If $n \ge k+1$, then*

$$R_C(n) = \sum_{j=0}^{n-2} f_C(j, n-j+1) p(j,t),$$

where $f_C(j,m) = 0$ for $j < 0$, $f_C(0,m) = 1$ for $m \ge 0$,

$$f_C(j,2) = \begin{cases} j+1, & 0 \le j \le k; \\ 0, & \textit{otherwise}, \end{cases}$$

$$f_C(j,m) = f_C(j-1,m) + f_C(j,m-1) - f_C(j-k,m-1) \textit{ for } m > 2.$$

Proof. The equation

$$R_C(n) = \sum_{j=0}^{n-2} f_C(j, n-j+1) p(j,t)$$

follows from the following observation:

Let exactly j components fail. Then the remaining $n-j$ good components, arranged on a line, form $n-j+1$ spaces, or cells. In order for the corresponding circular system to work, the j failed components should be distributed in those $n-j+1$ cells, so that all cells receive less than k failed components, and the first and last cells together should receive less than k failed component. This can be done in $f_C(j, n-j+1)$ ways.

Next we prove the recursive relation for $f_C(j,m)$'s, while the initial conditions are easily obtained. Consider the generating function:

$$\sum_{j=0}^{\infty} f_C(j,m) z^j = (1 + 2z + 3z^2 + \ldots + kz^{k-1})(1 + z + \ldots + z^{k-1})^{m-2}.$$

The factor $(1+z+\ldots+z^{k-1})^{m-2}$ accounts for the $m-2$ cells (except the first and the last) containing less than k objects. Let the first and last receive together i objects, where $0 \le i < k$. Then those i objects can be distributed in the two cells in $i+1$ ways, so we have the factor $1+2z+3z^2+\ldots+kz^{k-1}$. Thus

$$\begin{aligned}\sum_{j=0}^{\infty} f_C(j,m)z^j &= (1+2z+3z^2+\ldots+kz^{k-1})(1+z+\ldots+z^{k-1})^{m-2}\\ &= \left(\sum_{j=0}^{\infty} f_C(j,m-1)z^j\right)(1+z+\ldots+z^{k-1})^{m-2}\\ &= \frac{1-z^k}{1-z}\sum_{j=0}^{\infty} f_C(j,m-1)z^j\end{aligned}$$

and so

$$(1-z)\sum_{j=0}^{\infty} f_C(j,m)z^j = (1-z^k)\sum_{j=0}^{\infty} f_C(j,m-1)z^j.$$

By identifying the coefficients of z^j, we get

$$f_C(j,m) - f_C(j-1,m) = f_C(j,m-1) - f_C(j-k,m-1)$$

which is the same as the desired recursive relation. ■

Similarly, we have the result for the linear case.

THEOREM 5.5.2 *If* $n \ge k+1$, *then*

$$R_L(n,t) = \sum_{j=0}^{n} f_C(j,n-j+1)p(j,t)$$

where $f_L(j,m) = 0$ *for* $j < 0, f_L(0,m) = 1$ *for* $m \ge 0$,

$$f_L(j,1) = \begin{cases} 1, & 0 \le j < k;\\ 0, & \textit{otherwise}, \end{cases}$$

$$f_L(j,m) = f_L(j-1,m) + f_L(j,m-1) - f_L(j-k,m-1) \textit{ for } m \ge 1.$$

Proof. A similar argument as in the proof of Theorem 5.5.1 can be obtained by replacing "the first and last cells together should receive less than k failed components" with "every cell containing less than k failed components". Then the same recursive equation is obtained with different initial conditions. ■

Shanthikumar [236] used a different argument to get another recursive equation as follows. Recall that $N_L(d,n,k)$ is the number of ways that n nodes, including exactly d failed nodes, ordered in a line contains no k consecutive failed nodes, and $N(d,n,k)$ was solved in Theorem 2.3.3.

THEOREM 5.5.3 $R_L(n,t) = \sum_{d=0}^{n} N_L(d,n,k)p(d,t)$.

Proof. Since $T_1, T_2, \ldots, T_n$ are s-exchangeable, for every

$$\{T_1 \le t, T_2 \le t, \ldots, T_d \le t, T_{d+1} \ge t, \ldots, T_n \ge t\},$$

there are $N_L(d,n,k)$ ways to distribute the failed components. This gives the equation for $R_L(n)$. ■

Chapter 6

WINDOW SYSTEMS

An *m-window* is a sequence of m consecutive cells. We first identify a criterion to call a window "bad". For example, a window is bad if it contains k failed components, or a window is bad if it contains consecutive k failed components. The system failure is defined in terms of windows. It could be that the system fails if it contains a bad window. For example, the consecutive system we studied earlier defines a bad window as one which contains m failed components, called an *all-failed window*, and one bad window fails the system. But we can also define a system to be failed if every window is bad, or equivalently, the system succeeds if it contains a good window. In general, we can define a system to fail if it contains b bad windows, $1 \leq b \leq n - m + 1$ (or b consecutive bad windows), called a *b-fold-window system*.

We can recursively define a window system by using the last-obtained window system as the window in the new system. For example, a k-within-consecutive-m-out-of-n system is a k-window system where a window is bad if it contains k failed components. Then we can define another window system by using the k-within-consecutive-m-out-of-n system as an n-window.

6.1 THE k-WITHIN-CONSECUTIVE-m-OUT-OF-n SYSTEM

A *k-within-consecutive-m-out-of-n system*, or briefly a *(k, m, n) system*, is an n-line or n-cycle which fails if and only if there exist some m consecutive components (an *m-window*) containing at least k failed components. The problem of computing the reliability of a (k, m, n)

system can be viewed as the binomial version of the *generalized birthday problem* studied by Saperstin [223] as follows:

"Given a random assignment of w 1's and $n-w$ 0's, what is the probability that there is no set of m consecutive points in the arrangement containing k or more 1's?"

The case $k = 2$ was first studied by Naus [177] as an extension of the much quoted birthday problem which asks for the probability that no two birthdays of a group of w persons occur within m days given that no two fall on the same day.

Saperstein also considered the binomial version of the generalized birthday problem in which each 0 has a probability p, and each 1 has a probability q of occurring. By interpreting each 0 as a good component and each 1 as a failed component, then this problem is exactly the problem of computing the reliability of the (k, m, n) system.

Griffith [95] first proposed the (k, m, n) system as an extension of the consecutive system, i.e., the latter is the special case when $m = k$.

Two methods have been proposed to compute the reliability; each being efficient in a certain zone of (m, n).

Suppose $n = \ell m + r$, where $0 \le r < m$. First consider the case $r = 0$. Partition the n components consecutively into ℓ groups each with m components. Let d denote the total number of failed components, and $d_i < k$ the number of failed components in group i. Then $\sum_{i=1}^{\ell} d_i = d$. Note that the number of permutations satisfying $\{d_i\}$ is

$$\prod_{i=1}^{\ell} \binom{m}{d_i}.$$

Let $X_i(t)$ denote the number of failed components in the first t components of group i. Then the number of failed components in the m-sequence from $(i-1)m+t+1$ to $im+t$ can be written as $d_i - X_i(t) + X_{i+1}(t)$.

Let $R(d_1, d_2, \ldots, d_\ell)$ denote the reliability conditional on $d_1, d_2, \ldots, d_\ell$. Define

$$\alpha_i = \sum_{j=i}^{\ell-1}(k-d_j), \quad i = 1, 2, \ldots, \ell.$$

Then $\alpha_1 > \alpha_2 > \ldots > \alpha_\ell = 0$.

$$\begin{aligned} &R(d_1, d_2, \ldots, d_\ell) \\ &\quad = \Pr(\cap_{i=1}^{\ell-1} \cap_{t=1}^{m} (d_i - X_i(t) + X_{i+1}(t)) < k | \{d_i\}) \\ &\quad = \Pr(\cap_{t=1}^{m} (X_1(t) + \alpha_1 > X_2(t) + \alpha_2 > \ldots > X_\ell(t) + \alpha_\ell) | \{d_i\}). \end{aligned}$$

Note that $X_i(t) + \alpha_i$ describes a path, starting at α_i and ending at

$$X_i(m) + \alpha_i = d_i + \alpha_i.$$

We seek the probability that none of the ℓ paths intersect at any $t = 1, \ldots, m$.

Naus [178] gave a discrete version of the Karlin-McGregor Theorem [129] of coincidence probability.

LEMMA 6.1.1 *Consider ℓ one-dimensional (linearly ordered state space) Markov chains with identical transition probability $P(t)$ such that a state can either stay put or move to the next state. Let the ℓ labeled particles start at time 0 at $u_1, u_2, \ldots, u_\ell$ follow the ℓ Markov chains. Order the particles by $u_1 > u_2 > \ldots > u_\ell$. Then the probability that at time t, particle i is in state v_i, $v_1 > v_2 > \ldots > v_\ell$ without any two of the particles coinciding anywhere in $(0, t)$ is* $\det(P(t))$.

Proof. Each term in $\det(P(t))$ is

$$\Pr(\alpha_i \to d_{\pi(i)} + \alpha_{\pi(i)} : i = 1, 2, \ldots, \ell),$$

where $\{\pi(i)\}$ is a permutation of i. Consider a permutation π and a realization R in π such that two particles i and j cross their paths. Let $t_{ij}(R)$ be the first time they cross. Let R' be obtained from R by interchanging the paths of i and j at $t_{ij}(R)$. Then π', the permutation R' belongs to, is obtained from π by interchanging $\pi(i)$ and $\pi(j)$. Clearly, $\Pr(R) = \Pr(R')$. Further, R and R' have opposite signs in $\det(P(t))$ since π' differs from π by one interchange. So they cancel out in computing $\det(P(t))$. Therefore, the only realizations not cancelled out are those where paths do not intersect. Necessarily, $\pi(i) = i$. ∎

We restate Naus' results [178] for the generalized birthday problem in its binomial version to compute the reliability of the (k, m, n) system. Let the states be the numbers of failed components. Then for any states $u \leq v$,

$$P_{uv}(m) = \binom{m}{v - n} q^{v-u} p^{m-v+u}.$$

Let S_d denote a partition of d into $d_1, d_2, \ldots, d_\ell$ such that $d_i < k$ for all i. Set $u_i = \alpha_i$ and $v_i = \alpha_i + d_i$. By Lemma 6.1.1,

$$R(d_1, d_2, \ldots, d_\ell) = \det(P_{uv}(m)).$$

LEMMA 6.1.2 $R(n) = \sum_{d=0}^{n} \sum_{\{d_i\} \in S_d} \prod_{i=1}^{\ell} \binom{m}{d_i} \det(P_{uv}(m))$.

$P_{uv}(m)$ is an $\ell \times \ell$ matrix, and hence computing $\det(P_{uv}(m))$ takes $O(\ell^3)$ time (without using any advanced method). $|S_d|$ is the number of ways of partitioning d objects into ℓ groups without any k of them going to the same group. This number is exactly $N(d, \ell + d, k)$ which was given in Theorem 2.3.3. Therefore the computing of $R(m)$ takes $O(n\ell^3 N(k, d, \ell + d))$ time.

Hwang and Wright [117] proposed the transfer matrix method which was reported in Section 2.3 for the special case $m = k$. For the general m case, first assume $n/m = \ell$ is an integer. Again partition the n components consecutively into ℓ groups of m each. Define $L = \{\ell \in \{0,1\}^m : \ell \text{ contains fewer than } k \text{ 1's}\}$. Let M be an $|L| \times |L|$ matrix such that

$$M_{uv} = \begin{cases} 1, & \text{if } s_u \cup s_v \text{ is a failed line,} \\ 0, & \text{otherwise.} \end{cases}$$

Then

$$R(n) = q^n \Phi M^{\ell-1} \Phi^t,$$

where Φ is an $|L| \times 1$ vector with $\Phi_i = \rho^{d_i/2}$ for $1 \le i \le \ell$. This method requires $O(|L|^3 \log \ell)$ time. Since

$$|L| = \sum_{j=0}^{k-1} \binom{m}{j} < m^k,$$

this method is efficient when k is small and works for both lines and cycles.

Next suppose $r > 0$ and $\gcd(r, m) = c > 1$. Let $n' = n/c$ and $m' = m/c$. Partition the n components into n' groups of c each. Again, let d_i denote the number of failed components in group i, and define

$$S_d = \{\{d_i\} : d_i \le c, \sum_{i=j}^{j+m'-1} d_i < k\}.$$

Then using an argument similar to the one given by Wallenstein and Naus [249] for the continuous t case, we have

THEOREM 6.1.3 $R(n) = \sum_{d=0}^{n} \sum_{\{d_i\} \in S_d} \prod_{i=1}^{n'} \binom{c}{d_i} (\det(P_{uv}(c)))^{m'}$.

For $c = 1$, we can break m into two parts r and $m - r$. Suppose $X_i(r) = x_i$ for $1 \le i \le \ell + 1$. Then we compute the reliability as the product of two terms, one from u_i to x_i for $\ell + 1$ particles and the other from x_i to v_i for ℓ particles. We sum over all $\{x_i\}$ such that $\alpha_i \le x_i \le \min\{k, \alpha_i + t\}$.

For $n < 2m$, Sfakianakis, Kounias and Hillaris [233] gave a recursive equation.

THEOREM 6.1.4 *For* $n \leq 2m$,

$$R_L(k,m,n) = \sum_{x=0}^{k-1} R_L(k-x, n-m, 2n-2m)\binom{2m-n}{x} p^{2m-n-x} q^x.$$

Proof. The $2m-n$ components in the middle appear in every m-window. Suppose they contain x failed components and $2m-n-x$ good ones. By removing these $2m-n$ components, the system is reduced to a $(k-x)$-within-consecutive-$(n-m)$-out-of-$(2n-2m)$-system. ∎

They showed that Theorem 6.1.4 can be used to obtain closed form solutions when $n-m$ is small.

COROLLARY 6.1.5 $R_L(k,m,m+1) = \sum_{x=0}^{k-2} \binom{m-1}{x} p^{m-1-x} q^x + \binom{m-1}{k-1} p^{m-k+2} q^{k-1}$.

COROLLARY 6.1.6 $R_L(k,m,m+2) = \sum_{x=0}^{k-3} \binom{m-2}{x} p^{m-2-x} q^x + p^2(1+2q) \binom{m-2}{k-2} p^{m-k} q^{k-2} + \binom{m-2}{k-1} p^{m-k+3} q^{k-1}$.

COROLLARY 6.1.7 $R_L(k,m,m+3) = \sum_{x=0}^{k-4} \binom{m-3}{x} p^{m-3-x} q^x + (1-4q^3+3q^4) \binom{m-3}{k-3} p^{m-k} q^{k-3} + p^4(p^2+6q) \binom{m-3}{k-2} p^{m-k-1} q^{k-2} + p^6 \binom{m-3}{k-1} p^{m-k-2} q^{k-1}$.

Proof. Compute $R_L(3,3,6) = R_L(3,6)$ by using results for consecutive system, and compute $R_L(2,3,6)$ by using results in Section 6.2. ∎

Sfakianakis, Kounias and Hillaris [233] also gave two methods to obtain bounds. The first is to partition a k-within-consecutive-m-out-of-n system into $s = \lceil n/(r+3) \rceil$ subsystems with $r+3$ components each (except one of them may have fewer) and to obtain a lower bound by requiring all subsystems working. Define A_i to be the event that the subline $(i, i+m-1)$ is a failed line. In the second method they applied an improved Bonferoni inequality given in Kounias and Sotirakoglou [139] on $\Pr(\cup_{i-1}^{N} A_i)$, where $N \equiv n-m+1$, to obtain

THEOREM 6.1.8 $a_1 S_1 - a_2 S_2 + a_3 S_3 \leq \overline{R}_L(n) \leq \min\{1, S_1 - b_2 S_2 + b_3 S_3\}$, *where*
$a_1 = \frac{2N+t-1}{N(t+1)}$, $a_2 = \frac{2(N+2t-2)}{Nt(t+1)}$, $a_3 = \frac{6}{Nt(t+1)}$ *with* $t = \lfloor \frac{2((N-2)S_2 - 3S_3)}{(N-1)S_1 - 2S_2} \rfloor$;
$b_2 = \frac{2(2r-1)}{r(r+1)}$, $b_3 = \frac{6}{r(r+1)}$ *with* $r = \lfloor \frac{3S_3}{S_2} \rfloor + 2$;

$$S_1 = Nw,$$
$$S_2 = \binom{N-m+1}{2}w^2 + \sum_{u=1}^{\min\{m-1,N-1\}} \binom{N-u}{1}\Pr(A_iA_{i+u}),$$
$$S_3 = \binom{N-2m+2}{3}w^3 + 2w \sum_{u=1}^{\min\{m-1,N-m-1\}} \binom{N-u-m+1}{2}\Pr(A_iA_{i+u})+$$
$$\sum_{u=1}^{\min\{m-1,N-2\}} \sum_{v=1}^{\min\{m-1,N-1-u\}} \binom{N-u-v}{1}\Pr(A_iA_{i+u}A_{i+u+v}),$$

with

$$\Pr(A_i) = \sum_{x=k}^{m} q^x p^{m-x} = w,$$

$$\Pr(A_iA_j) = \begin{cases} w^2, & \text{if } j-i > m-1, \\ \displaystyle\sum_{x=k-j+i}^{m+i-j} \sum_{y=k-x}^{j-i} \sum_{z=k-x}^{j-i} \binom{m+i-j}{x} \\ \binom{j-i}{y}\binom{j-i}{z} q^{x+y+z} p^{m+j-i-x-y-z}, & \text{otherwise.} \end{cases}$$

$$\Pr(A_iA_jA_\ell) = \begin{cases} w^3, & \text{if } j-i > m-1, \ell-j > m-1, \\ \Pr(A_iA_j)w & \text{if } j-i \le m-1, \ell-j > m-1, \\ \Pr(A_jA_\ell)w & \text{if } j-i > m-1, \ell-j \le m-1, \\ \displaystyle\sum_{x=k-\ell+i}^{i+m-\ell} \sum_{y=k-x-j+i}^{\ell-j} \sum_{z=k-x-y}^{j-i} \sum_{u=k-x-y}^{j-i} \sum_{k-x-y}^{j-i} \\ \binom{i+m-\ell}{x}\binom{\ell-j}{y}\binom{j-i}{z}\binom{j-i}{u}\binom{\ell-j}{v} q^a p^{\ell+m-i-a}, \\ \text{where } a = x+y+z+u+v, & \text{otherwise.} \end{cases}$$

6.2 THE $(2, m, n)$ SYSTEM

Almost all results obtained for the consecutive system can be extended to the $(2, m, n)$ system.

Let $N_S(d, k, m, n)$ denote the number of the k-within-consecutive-m-out-of-n S-system which works, $S \in \{L, C\}$, containing exactly d failed components. Naus gave closed form solution for $k = 2$.

THEOREM 6.2.1 $N_L(d, 2, m, n) = \binom{n-(d-1)(m-1)}{d}$.

Proof. Consider the sequence of the d failed components. Then there must be at least $m-1$ good components between any two consecutive failed components. This fixes the positions of $(d-1)(m-1)$ good components. The remaining $n-d-(d-1)(m-1)$ good components can permute with the d failed components in any way. The number of such permutation is

$$\binom{n-(d-1)(m-1)}{d}.$$

■

Sfaniakakis, Kounias and Hilaris [233] obtained the corresponding number for circular case.

THEOREM 6.2.2 $R_C(d,2,m,n) = \frac{n}{n-d(m-1)}\binom{n-d(m-1)}{d}$.

Proof. Suppose that there is no failed component in the last $m-1$ nodes. Then the problem is reduced to $N_L(d,2,m,n-m+1) = \binom{n-d(m-1)}{d}$. If there is a failed component, then the $m-1$ nodes on each side must all be good and the problem is reduced to $N_L(d-1,2,m,n-2m+1) = \binom{n-d(m-1)-1}{d-1}$. There are $m-1$ such cases. Therefore

$$\begin{aligned} N_C(d,2,m,n) &= \binom{n-d(m-1)}{d} + (m-1)\binom{n-d(m+1)-1}{d-1} \\ &= \left(1+\frac{(m-1)d}{n-d(m-1)}\right)\binom{n-d(m-1)}{d} \\ &= \frac{n}{n-d(m-1)}\binom{n-d(m-1)}{d}. \end{aligned}$$

■

COROLLARY 6.2.3 *For the IID model and $k=2$,*

$$\begin{aligned} R_L(n) &= \sum_{d=0}^{n}\binom{n-(d-1)(m-1)}{d}q^d p^{n-d}, \\ R_C(n) &= \sum_{d=0}^{n}\frac{n}{n-d(m-1)}\binom{n-d(m-1)}{d}q^d p^{n-d}. \end{aligned}$$

Higashiyama, Ariyoshi and Kraetzl [98] gave a recursive equation under the IND model.

THEOREM 6.2.4 *For $k=2$ and $n \geq m$,*

$$R_L(1,n) = p_n R_L(1,n-1) + q_n\left(\sum_{i=n-m+1}^{n-1} p_i\right) R_L(1,n-m).$$

Proof. If the last component is good, then the problem is reduced to $R_L(1,n-1)$. If the last component is failed, the proceeding $m-1$ components must all be good. Then the problem is reduced to $R_L(1,n-m)$. ■

They also did the circular case.

THEOREM 6.2.5 *For $k = 2$,*

$$R_C(1,n) = \left(\prod_{i=n-m+2}^{n} p_i\right) R_L(1,n-m+1) + \sum_{i=1}^{m-1} q_{n-m+i+1} \left(\prod_{j=n-2m+i+2}^{n-m+i} p_j\right) \left(\prod_{j=n-m+i+2}^{n+i} p_j\right) R_L(i+1, n-2m+i+1).$$

Proof. The first term represents the reliability when the last $m-1$ components are all good, the second when component $n-m+i+1$, $1 \leq i \leq m-1$, is failed (and consequently, the $m-1$ preceding (succeeding) components are all good). ∎

In an unpublished manuscript, Sfakianakis proved

LEMMA 6.2.6 *For $k = 2$,*

$$I_i(1,n) = \frac{1}{p_i}\left(R(1,n) - R(1,i-m)\left(\prod_{j=i-m+1}^{i-1} p_j\right)\left(\prod_{j=i+1}^{i+m-1} p_j\right) R(i+m,n)\right).$$

Proof.

$$\begin{aligned}
&I_i(1,n) \\
&= R(p_1,\ldots,p_{i-1},1,p_{i+1},\ldots,p_n) - R(p_1,\ldots,p_{i-1},0,p_{i+1}\ldots,p_n) \\
&= \frac{1}{p_i}\Big(R(1,n) - q_i R(p_1,\ldots,p_{i-1},0,p_{i+1}\ldots,p_n)\Big) \\
&\quad -R(p_1,\ldots,p_{i-1},0,p_{i+1}\ldots,p_n) \\
&= \frac{1}{p_i}\Big(R(1,n) - R(p_1,\ldots,p_{i-1},0,p_{i+1}\ldots,p_n)\Big) \\
&= \frac{1}{p_i}\left(R(1,n) - R(1,i-m)\left(\prod_{j=i-m+1}^{i-1} p_j\right)\left(\prod_{j=i+1}^{i+m-1} p_j\right) R(i+m,n)\right),
\end{aligned}$$

where $p_j = 1$ for $j < 1$ or $j > n$. ∎

COROLLARY 6.2.7 *For the IID model, $I_i(n) = (R(n) - R(i-m)p^{x_i} R(n-i-m+1))/p$, where $x_i = \min\{m-1, i-1\} + \min\{m-1, n-i\}$.*

THEOREM 6.2.8 *$I_i^C \leq I_{i+1}^C$ for $1 \leq i \leq m-1$ for any k out of consecutive-m line under IND.*

Proof. Any m-sequence containing i also contains $i+1$. So a cutset containing i must also contain $i+1$. ∎

COROLLARY 6.2.9 *In an optimal assignment,* $p_i \le p_{i+1}$ *for* $1 \le i \le \min\{m-1, \lceil n/2 \rceil\}$.

Recently, Chang, Cui and Hwang [39] proved

THEOREM 6.2.10 *For* $k = 2$ *and* m *fixed,* $I_1^H \le I_i^H \le I_m^H$ *for all* i *under IID.*

Proof. True for $n \le 2m$ by Theorem 6.2.8. The general n case is proved by induction. Let $f_i(d,n)$ denote the number of d-cutsets containing i, where n is the total number of components. If the last component is working, then $f_i(d,n)$ is reduced to $f_i(d-1, n-m)$. Hence

$$f_i(d,n) = f_i(d, n-1) + f_i(d-1, n-m).$$

Theorem 6.2.10 now follows from induction. ■

COROLLARY 6.2.11 $I_1^B \le I_i^B \le I_m^B$ *for all* i.

THEOREM 6.2.12 *For* $k = 2$,m *fixed and* $n - i + 1 \ge 2m$,

$$I_i(1,n) = p_n I_i(1, n-1) + q_n \left(\prod_{j=n-m+1}^{n-1} p_j \right) I_i(1, n-m).$$

Proof.

$$\begin{aligned}
& p_n I_i(1,n) \\
&= R(1,n) - R(1, i-m) \left(\prod_{j=i-m+1}^{i-1} p_j \right) \left(\prod_{j=i+1}^{i+m-1} p_j \right) R(i+m, n) \\
&= p_n R(1, n-1) + q_n \left(\prod_{j=n-m+1}^{n-1} p_j \right) R(1, n-m) - \\
&\quad R(1, i-m) \left(\prod_{j=i-m+1}^{i-1} p_j \right) \left(\prod_{j=i+1}^{i+m-1} p_j \right) \\
&\quad \left(p_n R(i+m, n-1) + q_n \left(\prod_{j=n-m+1}^{n-1} p_j \right) R(i+m, n-m) \right) \\
&= p_n \left(R(1, n-1) - R(1, i-m) \left(\prod_{j=i-m+1}^{i-1} p_j \right) \left(\prod_{j=i+1}^{i+m-1} p_j \right) \right. \\
&\quad \left. R(i+m, n-1) \right) + q_n \left(\prod_{j=n-m+1}^{n-1} p_j \right) \left(R(1, m-n) - \right.
\end{aligned}$$

$$R(1,i-m)\left(\prod_{j=i+m+1}^{i-1} p_j\right)\left(\prod_{j=i+1}^{i+m-1} p_j\right)R(i+m,n-m)\Bigg)$$

$$= \quad p_nI_i(1,n-1)+q_n\left(\prod_{j=n-m+1}^{n-1} p_j\right)I_i(1,n-m).$$

■

COROLLARY 6.2.13 *For IID,* $I_i(n) = pI_i(n-1) + qI_i(n-m)$.

The optimal DLR dynamic assignment for the consecutive-2 system is still optimal for the 2-within-consecutive-m line with the modification that after a failed component, the $m-1$ currently best components should be assigned.

The following problems extended from the consecutive-2 system remain open:

1. Optimal 2-within-consecutive-m-out-of-n G system for $n > 2m$ ($n > 2m+1$ for cycle).
2. Optimal k-within-consecutive-m-out-of-n G circular system for $n \le 2m+1$.
3. Optimal 2-within-consecutive-m-out-of-n F system.

6.3 b-FOLD-WINDOW SYSTEMS

We first consider b-fold-nonoverlapping-consecutive system when a window is a consecutive system and the system fails if and only if there exist b nonoverlapping bad windows. Let $R_L^b(1,n)$ denote the reliability of such a (linear) system. Papastavridis [190] gave a recursive equation.

THEOREM 6.3.1 $R_L^b(1,n) = R_L^b(1,n-1) - \sum_{j=1}^{b} p_{n-jk}\left(\prod_{i=1}^{jk} q_{n-jk+i}\right)$ $\left(R_L^{b-j}(1,n-jk-1) - R_L^{b-j+1}(1,n-jk-1)\right)$.

Proof. When there are exactly jk failed components at the end for some $1 \le j \le b$, and the remaining $(n-jk-1)$-line contains exactly $b-j$ bad windows, then the n-line fails but the $(n-1)$-line works. The last term in ( ) represents the probability of exactly $b-j$ bad windows. ■

There are bn terms of R_L^b to compute, each requiring $O(b)$ time (the q terms can be separately computed). Hence $O(b^2n)$ time is needed.

Alevizos, Papastavridis and Sypass [5] did it for the circular model with a similar argument.

THEOREM 6.3.2 $R_C^b(1,n) = p_n R_L^b(1,n) + q_n R_C^b(1,n-1) - \sum_{j=1}^{b}\sum_{i=1}^{jk}\left(\prod_{\ell=n-jk+i}^{n+i-1} q_\ell\right)$
$p_{n-jk+i-1}p_{n+i}\Big(R_L^{b-j}(i+1, n-jk+i-2) - R_L^{b-j+1}(i+1, n-jk+i-2)\Big)$.

Proof. For the circular case, the set of jk consecutive failed components which covers component n can start anywhere from $n-jk+1$ to n. ∎

Since the computation of each R_C^b term requires $O(b^2k)$ time, the computation of $R_C^b(1,n)$ requires $O(b^3kn)$ time.

Let $N_L^b(d,k,n)$ denote the number of linear arrangements of n components which contain exactly d failed components but no b nonoverlapping subsequences of consecutive-k failed components. Papastarvidis proved

LEMMA 6.3.3 $R_L^b(d,k,n) = \binom{n-d+b}{b}\sum_{i\geq 0}(-1)^i\binom{n-d+1}{i}\binom{n-k(i+b)}{n-d}$.

Proof. There are $n-d+1$ spaces created by the working components. The number of ways of selecting b spaces is

$$\binom{n-d+1+b-1}{b} = \binom{n-d+b}{b}.$$

Insert k failed components into each of these b selected spaces. We still have $d-bk$ failed components to insert into the $n-d+1$ spaces without any k of them going into the same space; this number is exactly $N_L(d-bk, n-bk, k)$. Lemma 6.3.3 now follows from Theorem 2.3.3. ∎

THEOREM 6.3.4 $R_L^b(n) = \sum_{d=0}^{n} N_L^b(d,n,k)p^{n-d}q^d$.

LEMMA 6.3.5 $R_C^b(d,n,k) = \binom{n-d+b-1}{b}\binom{n-d}{d}\sum_{i\geq 0}(-1)^i\binom{n-d+1}{i}\binom{n-k(i+b)}{n-d}$.

Proof. Note that $n-d$ working components yield only $n-d$ spaces on a cycle. Hence the first term is

$$\binom{n-d+b-1}{b}.$$

The number of ways of inserting the remaining $n-bk$ failed components into $n-d$ spaces on a cycle without any of them going into the same space is $N_C(d-bk, k, n-d)$ which was given by Theorem 2.3.3. Lemma 6.3.5 follows immediately. ∎

THEOREM 6.3.6 $R^b_C(n) = \sum_{d=0}^{n} N^b_C(d,k,n)p^{n-d}q^d$.

All-bad windows are called *isolated* if there must exist a working component between any two windows. For example, for $k = 2$, four consecutive failed components consist of only one bad window since although the first two and the last two each continue a bad window, there is no working component separating them.

We next consider the b-isolated failed components with $f \geq k$, and the system fails if and only if there exist b-isolated bad windows. Let $R^{(b)}$ denote its reliability.

THEOREM 6.3.7 $R^{(b)}_L(1,n) = R^{(b)}_L(1,n-1) - \left(\prod_{i=n-k+1}^{n} q_i\right)p_{n-k}$ $\left(R^{(b-1)}_L(1,n-k-1) - R^{(b)}_L(1,n-k-1)\right)$.

Proof. The only case the $(n-1)$-line works but not the n-line is when exactly k components failed at the end of the line, and the first $n-k-1$ components contains exactly $b-1$ bad windows. ■

It takes $O(bn)$ time to compute $R^{(b)}_L$.

THEOREM 6.3.8 $R^{(b)}_C(1,n) = p_n R^{(b)}_L(1,n) + q_n R^{(b)}_C(1,n-1) -$ $\sum_{i=n-k+1}^{n}\left(\left(\prod_{j=i}^{i+k-1} q_j\right)p_{i-1}p_{i+k}\left(R^{b-1}_L(i+k+1,i-2) - R^b_L(i+k+1,i-2)\right)\right)$.

It takes $O(bkn)$ time to compute $R^{(b)}_C$.

Let $N^{(b)}(d,k,n)$ mean the same as $N^b(d,k,n)$ except for isolated windows.

LEMMA 6.3.9 $N^{(b)}_L(d,k,n) = \sum_{x=bk}^{d} \binom{n-d+b}{b}\binom{x-bk+1}{b-1} N_L(d-x,k,n-x-b)$.

Proof. The $n-d$ working components create $n-d+1$ spaces, exactly b of them contain at least k failed components each. Suppose these b spaces actually contain at least k failed components. Then the number of ways of partitioning the number of $x-bk$ into b spaces is same as the number of ways of partitioning the number $x-bk$ into b ordered parts, which is $\binom{x-bk+1}{b-1}$. Finally, the number of ways of distributing the $d-x$ failed components into the other $n-d+1-b$ spaces without any k of them going into one space is exactly $N(d-x,k,n-x-b)$, which was given in Theorem 2.3.3. ■

THEOREM 6.3.10 $R_L^{(b)}(d,k,n) = \sum_{d=0}^{n} N_L^{(b)}(d,k,n)p^{n-d}q^d.$

Similarly, we have

LEMMA 6.3.11 $N_C^{(b)}(d,k,n) = \sum_{x=b-k}^{d} \binom{n-d+b-1}{b}\binom{x-bk+1}{b-1} N_C(d-x,k,n-x-b).$

Proof. The $n-d$ working components create only $n-d$ spaces on a cycle. ∎

THEOREM 6.3.12 $R_C^{(b)}(d,k,n) = \sum_{d=0}^{n} N_C^{(b)}(d,k,n)p^{n-d}q^d.$

6.4 ASYMPTOTIC ANALYSIS

For a k-within-consecutive-m-out-of-n:F system, Papastavridis [189] gave the following asymptotic result.

THEOREM 6.4.1 *Let $m \geq k \geq 2$, and $q(t) = (t\lambda)^\alpha + o(t^\alpha)$ where α and λ are positive real constants. Then*

$$\lim_{n\to\infty} \Pr\{n^{1/k\alpha}T_n \leq t\} = 1 - \exp\left(-(t\lambda)^{k\alpha}\binom{m-1}{k-1}\right).$$

Proof. The proof of the theorem is based on Theorem 2, p399 of Barbour and Eagleson [13]. Consider the set

$$A = \{(i-j+1, i-j+2, \ldots, i) : 1 \leq i-j+1, i \leq n, k \leq j \leq m\}$$

which consists of j-tuples of numbers $1, 2, \ldots, n$ representing the component of the system in their ordering. Let $J = (i-j+1, \ldots, i) \in A$,

$$X_J = \begin{cases} 1, & \text{components } i-j+1 \text{ and } i \text{ are failed and there} \\ & \text{are } k \text{ failures among components } i-j+1, \ldots, i, \\ 0, & \text{otherwise,} \end{cases}$$

and $X = \sum_{J\in A} X_J$. It is clear that the system fails if and only if $X > 0$. Let $t_n = tn^{-1/k\alpha}$, where $t \geq 0$. We consider the system in the interval from 0 to t_n. Clearly,

$$p_J = E(X_J) = q(t_n)^k(1-q(t_n))^{j-k}\binom{j-2}{k-2}$$

and

$$E(X) = (n - m + 1)q(t_n)^k \sum_{j=k}^{m} (1 - q(t_n))^{j-k} \binom{j-2}{k-2} + \sum p_J,$$

where in the second summation J ranges over j-tuples of the numbers $1, 2, \ldots, m$, only. It is easy to see that

$$q(t_n)^k = (t\lambda)^{k\alpha}/n + o(1/n).$$

So,

$$\lim_{n\to\infty} E(X) = (t\lambda)^{k\alpha} \sum_{j=k}^{m} \binom{j-2}{k-2} = (t\lambda)^{k\alpha} \binom{m-1}{k-1}$$

because the summation $\sum p_J$ clearly goes to 0. By Theorem 2, p339 of Barbour and Eagleson [13],

$$|\Pr\{T_n \le t_n\} - (1 - \exp(-E(X)))|$$

is bounded above by

$$\min(1, 1/E(X))\Big(\sum p_J^2 + \sum (p_J p_K + E(X_J X_K))\Big),$$

where the first summation ranges over $J \in A$ and the second one ranges over $J, K \in A$ with $J \neq K$ and J and K have at least one common component. Clearly, we have

$$p_J p_K = (t\lambda)^{2k\alpha}/n^2 + o(1/n^2) \quad \text{for } J, K \in A.$$

And for $J, K \in A, J \neq K$ and having at least one common component, we have

$$E(X_J X_k) = O(n^{-1-1/k}).$$

Therefore, the above mentioned upper bound goes to 0. The theorem then follows. ■

Next we study the b-fold-nonoverlapping consecutive system. Poisson convergence is used to approximate some random numbers with small component failure probability in system. Usually one needs to compute some moments of the process by using Poisson convergence, but Fu [80] provided a new elementary and direct method. In the following, his method is presented.

A linear connected system is called b-Λ^* linearly connected when it fails if, and only if, there are b or more non-overlapping Λ^* patterns have occurred. A consecutive k-out-of-n:F system is the case where $b = 1$, and the pattern $\Lambda^* = F \ldots F$ is consecutive k or more failures. The study of asymptotic analysis for b-fold-consecutive k-out-of-n:F system here is to discuss the reliability of a large b-Λ^* linearly connected system when the failure probabilities of components are very small. Since we assume that the components operate independently, the reliability of the system is equal to the probability that there are fewer than b non-overlapping Λ^* patterns which occur in tossing a coin n times. In order to get results, it is sufficient to study the distribution of the number of non-overlapping Λ^* patterns which occur in coin tossing.

Let w be an n-tuples as a realization of tossing a coin n times, and $N_{n,k}(w)$ be the number of non-overlapping failure runs of size k. For example, with $n = 8, k = 2$, and $w = (SFFSFFFF)$, then $N_{8,2}(w) = 3$. For the same w, we have $N_{8,3}(w) = 1$ and $N_{8,4}(w) = 0$. Therefore, the reliability of a b-fold-consecutive k-out-of-n:F system is equal to the probability of $N_{n,k} \leq m - 1$ (omitting w means true for all w).

It is well-known that if the failure probabilities of components are very small and n is large, then the random variable $N_{n,k}$ converges to a Poisson random variable. For our problem, we need to find the convergence conditions and the Poisson distribution parameter.

THEOREM 6.4.2 *For a given k, if the failure probability of a Bernoulli trial (independently) is $q_{in} = q_n$, for $1 \leq i \leq n$, and satisfies the condition*

$$\lim_{n\to\infty} nq_n^k \to \lambda,$$

then the random variable $N_{n,k}$ converges to a Poisson random variable with mean λ; i.e., for every $x = 0, 1, \ldots,$

$$\lim_{n\to\infty} \Pr\{N_{n,k} = x\} = e^{-\lambda}\lambda^x/x!.$$

In order to prove the theorem, we need the following two lemmas whose proofs are omitted.

LEMMA 6.4.3 *For given k, m, n with $0 \leq m \leq n$, we have*

$$(1 - q_n^k)^m \leq \Pr(N_{m,k}(w) = 0) \leq (1 - q_n^k + q_n^{k+1})^{m-k+1}.$$

LEMMA 6.4.4 *For any two fixed non-negative integers c_1 and c_2,*

$$\lim_{n\to\infty} \binom{n - c_1}{x} (1 - \lambda/n + o(1/n))^{n-c_2} (\lambda/n + o(1/n))^x = e^{-\lambda}\lambda^x/x!.$$

Proof of Theorem 6.4.2 For k, n and x, define

$$A = A(x, n, k) = \{w : N_{n,k}(w) = x\}.$$

In order to prove the theorem, we shall construct two sets A^* and A_* satisfying the following three properties.

(i) $A_* \subseteq A \subseteq A^*$ and so $\Pr\{A_*\} \leq \Pr\{A\} \leq \Pr\{A^*\}$.

(ii) A lower bound of $\Pr(A_*)$ and an upper bound of $\Pr(A^*)$ can be easily obtained.

(ii) Both bounds converge to the same limit.

A vital property of coin tossing important to the proof is that, at any time t, the occurrence of a pattern Λ^* in the future is independent of the past.

Let $x \geq 1$. Select x positions randomly from the $n - xk + x$ positions and put the $\Lambda^* = \underbrace{F \dots F}_{k}$ patterns into the x positions respectively. Denote by $t_1^*, \dots, t_{x+1}^*$ the numbers of positions between the selected x of Λ^* patterns respectively (including the number of positions preceding the first selected pattern Λ^* and the number of positions after the last selected pattern Λ^*). It follows that $t_i^*, i = 1, \dots, x+1$, are integers, $0 \leq t_i^*$ and $\sum_{i=1}^{x+1} t_i^* = n - xk$. Let

$$\Gamma_x^* = \{\mathbf{t}^* = (t_1^*, \dots, t_{x+1}^*) : \sum_{i=1}^{x+1} t_i^* = n - xk, t_i^* \geq 0\}.$$

Note that the total number of $\mathbf{t}^*$'s in Γ_x^* is $\binom{n-(k-1)x}{x}$. For every $\mathbf{t}^* = (t_1^*, \dots, t_{x+1}^*) \in \Gamma_x^*$, define

$$A^*(x, n, k, t^*) = \{w = (SFS \dots F\Lambda^*, \dots, \Lambda^* S \dots F) : \text{no } \Lambda^* \text{ pattern has occurred in all the subsequences } (\underbrace{SS \dots FS}_{t_i^*}), i = 1, \dots, x+1\};$$

and their union

$$A^* = A^*(x, n, k) = \bigcup_{\mathbf{t}^* \in \Gamma_x^*} A^*(x, n, k, \mathbf{t}^*).$$

Next, define an auxiliary pattern $\Lambda = S\Lambda^*$, i.e., an S followed by consecutive k failures. Similarly, select x positions randomly from the $n - x(k-1) - x = n - xk$ positions and put x of the Λ patterns into the positions. Denote by $t_1, \dots, t_{x+1}$ the numbers of positions between the patterns respectively. It follows that $t_i, i = 1, \dots, x+1$, are integers satisfying $0 \leq t_i \leq n - x(k+1)$ and $\sum_{i=1}^{x+1} t_i = n - x(k+1)$. Let

$$\Gamma_x = \{\mathbf{t} = (t_1, \dots, t_{x+1}) : \sum_{i=1}^{x+1} t_i = n - x(k+1), t_i \geq 0\}.$$

Note that the total numbers of $\mathbf{t}$ in Γ_x is $\binom{n-kx}{x}$. For every $\mathbf{t} \in \Gamma_x$, define

$$A_*(x,n,k,\mathbf{t}) = \{w = (SFS \dots F\Lambda, \dots, \Lambda S \dots F) : \text{no } \Lambda \text{ pattern has occurred in all the subsequences } (\underbrace{SS \dots FS}_{t_i}), i = 1, \dots, x+1\};$$

and their union

$$A_* = A_*(x,n,k) = \bigcup_{\mathbf{t} \in \Gamma_x} A_*(x,n,k,\mathbf{t}).$$

Note that for all $\mathbf{t}, \mathbf{t}' \in \Gamma_x$ with $\mathbf{t} \neq \mathbf{t}'$,

$$A_*(x,n,k,\mathbf{t}) \cap A_*(x,n,k,\mathbf{t}') = \phi.$$

It follows from the definitions of $A^*(x,n,k)$ and $A_*(x,n,k)$ that

$$A_*(x,n,k) \subseteq A(x,n,k) \subseteq A^*(x,n,k).$$

Hence

$$\Pr\{A_*(x,n,k)\} \le \Pr\{A(x,n,k)\} \le \Pr\{A^*(x,n,k)\}.$$

According to Lemma 6.4.3, for $x = 0, 1, \dots,$

$$\begin{aligned}
\Pr\{A^*(x,n,k)\} &= \Pr\{\bigcup_{\mathbf{t}^* \in \Gamma_x^*} A^*(x,n,k,\mathbf{t}^*)\} \\
&\le \sum_{\mathbf{t}^* \in \Gamma_x^*} \Pr\{A^*(x,n,k,\mathbf{t}^*)\} \\
&\le \sum_{\mathbf{t}^* \in \Gamma_x^*} (1 - q_n^k + q_n^{k+1})^{\sum_{i=1}^{x+1} t_i^* - k + 1} (q_n^k)^x \\
&= \binom{n-(k-1)x}{x} (1 - q_n^k + q_n^{k+1})^{n-x(k+1)-k+1} (q_n^k)^x.
\end{aligned}$$

Similarly, for given x it follows from the definition of $A_*(x,n,k,\mathbf{t})$ that

$$\begin{aligned}
\Pr\{A_*(x,n,k)\} &= \Pr\{\bigcup_{\mathbf{t} \in \Gamma_x} A_*(x,n,k,\mathbf{t})\} \\
&\ge \sum_{\mathbf{t} \in \Gamma_x} \Pr\{A_*(x,n,k,\mathbf{t})\} \\
&\ge \sum_{\mathbf{t} \in \Gamma_x} (1 - q_n^k)^{\sum_{i=1}^{x+1} t_i} (p_n q_n^k)^x \\
&= \binom{n-kx}{x} (1 - q_n^k)^{n-x(k+1)} (q_n^k)^x p_n^x.
\end{aligned}$$

By the condition that $\lim_{n\to\infty} nq_n^k = \lambda$, we have

$$\begin{aligned}
&\binom{n-xk}{x}(1-\lambda/n+o(1/n))^{n-x(k+1)}(\lambda/n+o(1/n))^x(1-(\lambda/n)^{1/k})^x \\
\le\ & \Pr\{N_{n,k}=x\} \\
\le\ & \binom{n-(k-1)x}{x}(1-\lambda/n+o(1/n))^{n-x(k+1)-k+1}(\lambda/n+o(1/n))^x.
\end{aligned}$$

Taking $n \to \infty$, the theorem follows immediately from Lemma 6.4.4. ∎

The above Fu's proof [80] of this theorem is more direct and elementary than using the Bonferroni inequalities and Stein–Chen method. The proof needs no computation of moments of any order. It is almost as elementary as the proof that a binomial random variable tends to a Poisson random variable under the condition $nq^n \to \lambda$ as $n \to \infty$.

Three theorems related to the above result are given in the following without proofs.

THEOREM 6.4.5 *For a b-fold-nonoverlapping consecutive-k system with equal failure probability for every component. The system reliability converges to a Poisson random variable with mean λ if and only if* $\lim_{n\to\infty} nq_n = \lambda > 0$.

THEOREM 6.4.6 *For a b-fold-nonoverlapping consecutive-k system with independently unequal failure probabilities for components. The system reliability converges to a Poisson random variable with mean λ if and only if* $\lim_{n\to\infty} \sum_{i=1}^{n-k+1} u_{in} = \lambda > 0$ *and* $\lim_{n\to\infty} \sum_{i=1}^{n-k+1} u_{in}^2 = 0$, *where* $u_{in} = \prod_{j=0}^{k-1} q_{i+j,n}$ *for* $i = 1, \ldots, n-k+1$.

THEOREM 6.4.7 *For a b-fold-nonoverlapping consecutive-k system where the i-th component has reliability $p_i = 1 - q_i$ for $i = 1, \ldots, n$, and the components operate independently, the system reliability $R_L(b, n, k; p_i)$ satisfies*

$$|R_L(b,n,k;p_i) - \sum_{x=0}^{b-1} e^{-\lambda_n}\lambda_n^x/x!| \le (2k+2+nq_*^k)q_*^k,$$

where $q_* = \max_{i\ge 1} q_i$ *and* $\lambda_n = \sum_{i=k+1}^{n} p_{i-k}q_{i-k+1}\cdots q_i$.

Theorem 6.4.7 was obtained by Godbole [88]. It is useful for getting the limit of $R_L(b, n, k; p_i)$ under some conditions, such as $q_* = \frac{\lambda}{n^{1/k}}$ with $\lambda > 0$ and $\lim_{n\to\infty} \lambda_n$ exists.

Chapter 7

THE NETWORK MODEL

In many applications like the oil pipe-line or the telecommunication system, the mission of the system is to transmit something from a source to a sink. The consecutive-k condition can be interpreted as that a node is connected by links to the next k nodes, and the transmission will fail at the node if and only if the next k nodes all fail. Two assumptions are tacitly made in such an interpretation. The first is that links never fail. The second is that the first and the last node always work for either failure would fail the transmission mission. Therefore it is convenient not to count the first node, called the *source*, and the last node, called the *sink*, as part of the consecutive system.

In the network model, we relax on the first assumption by allowing link failures. More precisely, the system $S(0, n+1)$ is represented by a digraph with $n+2$ vertices $(0, 1, \ldots, n, n+1)$ and links from node i to j for $0 \leq i < j \leq n+1$ and $j - i \leq k$. We refer to node 0 as source and node $n+1$ as sink. In the system, node i has reliability p_i and a link from i to j has reliability p_{ij} (for the IID model, the system is denoted by $S(n)$ with reliability $R(n)$, p_i is reduced p and p_{ij} to p_{j-i}). We can also treat the ordinary system as the network system where $p_{ij} = 1$ for all i, j (and without source or sink). Only the consecutive-2 network system has been studied in the literature.

7.1 THE LINEAR CONSECUTIVE-2 NETWORK SYSTEM

Chen, Hwang and Li [56] gave a recursive equation for the IID model. They also commented that it is straightforward to extend it to the IND model. We do the IND model here.

THEOREM 7.1.1 *$R(0,1) = p_{01}$, $R(0,2) = p_{02} + q_{02}p_{01}p_1p_{12}$, $R(0,3) = p_{01}p_1p_{13} + p_{02}p_2p_{23} - p_{01}p_1p_{13}p_{02}p_2p_{23} + p_{01}p_1p_{12}p_2p_{23}q_{02}q_{13}$. For $n \geq 3$,*

$$\begin{aligned} R(0,n+1) &= [p_1p_{01}(1-q_{02}q_{12})/p_{12}]R(1,n+1) + \\ &\quad (1-p_1p_{01})p_2p_{02}R(2,n+1) - \\ &\quad (p_1p_{01}p_3p_{13}q_2p_{02}/p_{12})R(3,n+1). \end{aligned}$$

Proof. $R(0,1), R(0,2)$ and $R(0,3)$ are easily verified. Define $R'(0,n+1)$ to be the reliability of the system when the first two nodes are both sources. Namely, $R'(0,n+1)$ can be obtained from $R(0,n+1)$ by setting $p_1 = p_{01} = 1$. We give a recursive equation of $R(0,n+1)$ which also involves the R' function. By considering the three cases: both nodes 1 and 2 working, node 1 working but node 2 failed, node 1 failed but node 2 working, we have for $n \geq 3$,

$$\begin{aligned} R(0,n+1) &= p_1p_2\{p_{01}[(1-q_{02}q_{12})R'(1,n+1) + \\ &\quad q_{02}q_{12}p_{13}p_3R(3,n+1)] + q_{01}p_{02}R(2,n+1)\} + \\ &\quad p_1q_2p_{01}p_{13}p_3R(3,n+1) + q_1p_2p_{02}R(2,n+1) \\ &= p_1p_{01}p_2(1-q_{02}q_{12})R'(1,n+1) + \\ &\quad p_{02}p_2(1-p_1p_{01})R(2,n+1) + \\ &\quad p_1p_{01}p_3p_{13}(p_2q_{02}q_{12}+q_2)R(3,n+1). \end{aligned} \tag{7.1.1}$$

Also

$$\begin{aligned} R'(0,n+1) &= p_2(1-q_{02}q_{12})R'(1,n+1) + \\ &\quad p_3p_{13}(p_2q_{02}q_{12}+q_2)R(3,n+1). \end{aligned} \tag{7.1.2}$$

Substituting (7.1.2) into (7.1.1), we obtain

$$R(0,n+1) = p_1p_{01}R'(0,n+1) + p_{02}p_2(1-p_2p_{12})R(2,n+1)$$

or

$$R(1,n+1) = p_2p_{12}R'(1,n+1) + p_{13}p_3(1-p_2p_{12})R(3,n+1). \tag{7.1.3}$$

Substituting $R'(1,n+1)$ of (7.1.3) back to (7.1.1),

$$\begin{aligned} R(0,n+1) &= [p_1p_{01}(1-q_{02}q_{12})/p_{12}][R(1,n+1) - p_{13}p_3 \\ &\quad (1-p_2p_{12}R(3,n+1)] + p_2p_{02}(1-p_1p_{01} \\ &\quad R(2,n+1) + p_1p_{01}p_3p_{13}(p_2q_{02}q_{12}+q_2)R(3,n+1) \\ &= [p_1p_{01}(1-q_{02}q_{12})/p_{12}]R(1,n+1) + p_2p_{02}(1-p_1p_{01}) \\ &\quad R(2,n+1) - (p_1p_{01}p_3p_{13}q_2p_{02}/p_{12})R(3,n+1). \end{aligned}$$

■

COROLLARY 7.1.2 *For the IID model,* $R(0) = p_1$, $R(1) = p_2 + q_2pp_1^2$, $R(2) = (2 - p_2pp_1)p_1pp_2 + p^2p_1^3q_2^2$, *and for* $n \geq 3$,

$$\begin{aligned} R(n) &= p(1 - q_1q_2)R(n-1) + pp_2(1 - pp_1)R(n-2) - \\ & \quad p^2qp_2^2R(n-3). \end{aligned} \tag{7.1.4}$$

Clearly, $R(0, n+1)$ and $R(n)$ can be computed in linear time. Let $G(t)$ denote the generating function of $R(n)$.

THEOREM 7.1.3 $G(t) = \dfrac{p_1 + p_2(1 - pp_1)t - pq_1p_2^2t^2}{1 - p(1 - q_1q_2)t - pp_2(1 - pp_1)t^2 + p^2q_1p_2^2t^3}$.

Proof. Multiplying (7.1.4) by t^n and summing over $n \geq 3$.

$$G(t) - p_1 - (p_2 + q_2pp_1)t - pp_1[2p_2 - pp_1(p_2^2 - p_1q_2^2)]t^2 = p(1 - q_1q_2)t$$
$$[G(t) - p_1 - (p_2 + q_2pp_1)t] + pp_2(1 - pp_1)t^2[G(t) - p_1] - p^2q_1p_2^2t^3G(t).$$

Solving for $G(t)$, we obtain Theorem 7.1.4. ■

The recurrence equation (7.1.4) has characteristic polynomial

$$f(x) = x^3 - p(1 - q_1q_2)x^2 - pp_2(1 - pp_1)x + p^2q_1p_2^2. \tag{7.1.5}$$

Since

$$\begin{aligned} f(1) &= 1 - p(1 - q_1q_2) - pp_2(1 - pp_1) + p^2q_1p_2^2 \\ &= [1 - p(1 - q_1q_2)](1 - pp_2) > 0, \\ f(p(1 - q_1q_2)) &= -pp_2(1 - pp_1)p(1 - q_1q_2) + p^2q_1p_2^2 \\ &= -p^2p_2(1 - pp_1 - q_1q_2 + pp_1q_1q_2 - q_1p_2) \\ &= -p^2p_1p_2[1 - p(1 - q_1q_2)] < 0, \\ f(pp_2q_1) &= p^3p_2^3q_1^3 - p(1 - q_1q_2)p^2p_2^2q_1^2 - \\ & \quad pp_2(1 - pp_1)pp_2q_1 + p^2q_3p_2^2 \\ &= p^3q_1p_2^2(1 - q_1)^2 \geq 0, \end{aligned}$$

f has two roots α and β satisfying

$$1 > \alpha > p(1 - q_1q_2) > \beta \geq pp_2q_1. \tag{7.1.6}$$

The third root γ is negative since $\alpha\beta\gamma = -p^2q_1p_2^2 \leq 0$. On the other hand, $\alpha + \beta + \gamma = p(1 - q_1q_2)$ and (7.1.6) yield

$$\alpha + \gamma = p(1 - q_1q_2) - \beta > 0,$$
$$\beta + \gamma = p(1 - q_1q_2) - \alpha < 0.$$

Thus we have

LEMMA 7.1.4 $1 > \alpha > p(1-q_1q_2) > \beta \geq pp_2q_1 \geq 0 > -\beta > \gamma > -\alpha > -1$.

THEOREM 7.1.5 *For* $0 < p(1-q_1q_2) < 1$,

$$R(n) = a\alpha^n + b\beta^n + c\gamma^n,$$

where α, β, γ *are the three roots of* $f(x)$ *satisfying* (7.1.6) *and*

$$\begin{aligned} a &= \alpha^2(pp_2q_1 - \alpha)/[p(\alpha-\beta)(\gamma-\alpha)] > 0, \\ b &= \beta^2(pp_2q_1 - \beta)/[p(\beta-\gamma)(\alpha-\beta)] < 0, \\ c &= \gamma^2(pp_2q_1 - \gamma)/[p(\gamma-\alpha)(\beta-\gamma)] < 0. \end{aligned}$$

Proof. a, b, c are solved by using the initial conditions $R(0), R(1), R(2)$ and Cramer's rule. ■

Write $R(n) = R(n; p, p_1, p_2)$. Suppose $r > s > t$. The question is which mapping from $\{r, s, t\}$ to $\{p, p_1, p_2\}$ maximizes $R(n; p, p_1, p_2)$?

THEOREM 7.1.6 $R(n; r, t, s)$ *maximizes* $R(n; p, p_1, p_2)$ *for* n *large, where* $\{p, p_1, p_2\} = \{r, s, t\}$.

Proof. Let $\{u, v, w\} = \{r, s, t\}$ and $u > v$. It suffices to prove (i) $R(n; w, u, v) \leq R(n; w, v, u)$, (ii) $R(n; u, w, v) \geq R(n; v, w, u)$.

To prove (i), define $f(x)$ by setting $(p, p_1, p_2) = (w, u, v)$ and $\bar{f}(x)$ by setting $(p, p_1, p_2) = (w, v, u)$. Suppose $\bar{\alpha}$, $\bar{\beta}$, $\bar{\gamma}$ are the three roots of $\bar{f}(x)$. It is easily verified

$$0 < w[1-(1-u)(1-v)] < 1.$$

By Lemma 7.1.4,

$$1 > \bar{\alpha} > p[1-(1-u)(1-v)].$$

Hence

$$\begin{aligned} f(x) - \bar{f}(x) &= [wu(1-Nv) - wv(1-wu)]x + \\ &\quad w^2[(1-u)v^2 - (1-v)u^2] \\ &= w(u-v)\{x - w[1-(1-u)(1-v)]\} \\ &> 0 \qquad \text{for } x > w[1-(1-u)(1-v)]. \end{aligned}$$

In particular, $f(\bar{\alpha}) > \bar{f}(\bar{\alpha}) = 0$. It follows $\alpha > \bar{\alpha}$. For n large

$$R(n; w, u, v) \to a\alpha^n > R(n; w, v, u) \to \bar{a}\bar{\alpha}^n.$$

To prove (ii), define $f(x)$ by setting $(p, p_1, p_2) = (u, w, v)$ and $\bar{f}(x)$ by setting $(p, p_1, p_2) = (v, w, u)$. By Lemma 7.1.4,

$$1 > \alpha > u[1 - w(1 - v)] > uv,$$

$$1 > \bar{\alpha} > v[1 - w(1 - u)] > uv.$$

Hence

$$\begin{aligned} f(x) - \bar{f}(x) &= [v(w + u - wu) - u(w + v - wv)]x^2 + \\ &\quad uv[(1 - wv) - (1 - wu)]x \\ &= w(v - u)x(x - uv) < 0 \quad \text{for } x > uv. \end{aligned}$$

Hence $\alpha > \bar{\alpha}$ and Theorem 7.1.6 follows. ∎

7.2 CONNECTIVITY AND HAMILTONIAN RELIABILITY

For circular system, the concept of source and sink does not apply. Hwang and Yao [119] captured the spirit of the flow model (in which a flow exists without requiring all working nodes to participate) to set the condition "a flow from a node back to itself exists" for the cycle, or equivalently, "a circuit consisting of only working nodes and links exists." We call this the *flow reliability* (FR). Two other kinds of reliability were suggested by Hwang and Li [111] (they also work for links).

Connectivity reliability (CR). The set of working nodes is strongly connected and nontrivial (meaning there are at least two working nodes). For a line, the requirement is that the source can reach every working node, and every working node can reach the sink.

Hamiltonian reliability (HR). There exists a hamiltonian circuit (path for the line) among the working nodes.

Hwang and Li [111] gave characterizations of a working system under the two notions of reliability. For a state s let $G(s)$ denote the digraph of working nodes, and links between working nodes. A node of $G(s)$ is called an *island* if it has either no inlink or no outlink, except the source and the sink can have neither inlink nor outlink.

THEOREM 7.2.1 *s is a failed state under CR if and only if $G(s)$ contains either an island or two circuits not strongly connected.*

Proof. The "if" part is obvious. To prove the "only if" part, suppose that s is a failed state but $G(s)$ contains no island.

(a) Linear system: Let i be the first node in $G(s)$ the source can't reach. Since i is not an island, either link $(i-2, i)$ and node $i-2$ or link

$(i-1, i)$ and node $i-1$ are working. So i can be reached. Similarly, we can show that there does not exist a last node which cannot reach the sink.

(b) Circular system: Suppose that $G(s)$ contains two nodes i and j such that i cannot reach j. Let I be the set of nodes i can reach, and J the set of nodes which can reach j. Then I contains a circuit I' and J a circuit J', while I' cannot reach J'. ■

LEMMA 7.2.2 (i) *Suppose $G(s)$ is connected and contains neither node i nor link $(i-1, i+1)$ for some i. Then $G(s)$ is empty.*

(ii) *Suppose a hamiltonian circuit H' contains both node i and link $(i-1, i+1)$ for some i. Then H contains every node and every 2-link, plus n is odd.*

Proof. (i) Node $i+1$ is an island since it has no inlink. Hence node $i+1$ is not in $G(s)$. Furthermore, link $(i, i+2)$ is not in $G(s)$ since i is not. Similarly, we can argue that node $i-1$ and link $(i-2, i)$ are not in $G(s)$. Repeating this argument, then node j and link $(j-1, j+1)$ for every j are not in $G(s)$. Namely $G(s)$ contains no node.

(ii) The existence of link $(i-1, i+1)$ implies that nodes $i-1$ and $i+1$ are in H. Since each node in H has exactly one inlink and one outlink, links $(i-2, i)$ and $(i, i+2)$ are in H. Thus the containment of node i and link $(i-1, i+1)$ forces the containment of node $i-1$ and link $(i-2, i)$, as well as node $i+1$ and link $(i, i+2)$. Repeating this argument, H contains node j and link $(j-1, j+1)$ for every j. Namely, H contains all nodes and all 2-links. Since H is connected, n must be odd. ■

THEOREM 7.2.3 *s is a working state under HR if and only if it is so under CR and either* (i) *i and $i+1$ in $G(s)$ implies $i \to i+1$ in $G(s)$ or* (ii) *in the circular case, $G(s)$ contains all nodes and all 2-links, plus n is odd.*

Proof. "If": Clearly, (ii) induces a hamiltonian circuit. To prove (i), let $i \in G(s)$. If $i+1 \in G(s)$, then $i \to i+1 \in G(s)$. If $i+1 \notin G(s)$, by Lemma 7.2.2 (i), $i \to i+2 \in G(s)$. Therefore i can always reach the next node in $G(s)$. Clearly, $G(s)$ contains a hamiltonian circuit.

"Only if": Suppose that i and $i+1$ are in $G(s)$ but not $i \to i+1$. In the linear case it is easily seen that s is not a working state since i can neither precede nor succeed $i+1$ in a hamiltonian path. In the circular case, a hamiltonian circuit H must contain $i-1 \to i+1$. By Lemma 7.2.2 (ii), the existence of node i and link $i-1 \to i+1$ forces H to contain all nodes and all 2-links, plus n is odd. ■

Next we give a linear time algorithm to compute $HR_L(n)$.

THEOREM 7.2.4 *For* $k = 2$, $\mathrm{HR}_L(0,1) = p_{01}$, $\mathrm{HR}_L(0,2) = p_1p_{01}p_{12} + q_1p_{02}$, *and* $\mathrm{HR}_L(0,n+1) = p_1p_{01}\mathrm{HR}_L(1,n+1) + q_1p_{02}p_2\mathrm{HR}_L(2,n+1)$ *for* $n \geq 2$.

COROLLARY 7.2.5 *For IID model we have* $\mathrm{HR}_L(0) = p_1$, $\mathrm{HR}_L(1) = pp_1^2 + qp_2$, *and for* $n \geq 2$,

$$\mathrm{HR}_L(n) = pp_1\mathrm{HR}_L(n-1) + qpp_2\mathrm{HR}_L(n-2). \qquad (7.2.1)$$

Let $G(t)$ denote the generating function of $\mathrm{HR}_L(n)$.

THEOREM 7.2.6 $G(t) = (p_1 + qp_2t)/(1 - pp_1t - qpp_2t^2)$.

Proof. $G(t) - p_1 - (pp_1^2 + qp_2)t = pp_1t(G(t) - p_1) + qpp_2t^2G(t)$. Solving for $G(t)$, we obtain Theorem 7.2.6. ∎

THEOREM 7.2.7 $\mathrm{HR}_L(n) = a\alpha^n + b\beta^n$, *where*

$$a = \frac{pp_1 + qp_2 - p_1\beta}{\alpha - \beta}, \quad b = \frac{p_1\alpha - pp_1 - qp_2}{\alpha - \beta},$$

$$\alpha = \frac{pp_1 + \sqrt{(pp_1)^2 + 4qpp_2}}{2}, \quad \beta = \frac{pp_1 - \sqrt{(pp_1)^2 + 4qpp_2}}{2}. \qquad (7.2.2)$$

Proof. (7.2.1) has the characteristic polynomial

$$f(x) = x^2 - pp_1x - qpp_2.$$

The roots of $f(x)$ are α and β as shown in (7.2.2). write

$$\mathrm{HR}_L(n) = a\alpha^n + b\beta^n.$$

Then $\mathrm{HR}_L(0) = a + b = p_1$, $\mathrm{HR}_L(1) = a\alpha + b\alpha = pp_1 + qp_2$. Solving for a and b we obtain (7.2.2). ∎

Note that $\mathrm{HR}_L(n)$ given by Theorem 7.2.7 can be computed in $O(\log n)$ time.

THEOREM 7.2.8 $\mathrm{HR}_C(n) = \sum_{d=0}^{\lfloor n/2 \rfloor} \frac{n}{n-d}\binom{n-d}{d}(qp_2)^d p^{n-d} p_1^{n-2d} + \epsilon_n (pp_2^2)^n (1 - p_1^n)$, *where* $\epsilon_n = 1$ *if* n *is odd and* $= 0$ *if* n *is even.*

Proof. By the Kaplansky cycle theorem,

$$\frac{n}{n-d}\binom{n-d}{d}$$

is the number of cycles containing exactly d failed nodes without two of them consecutive. By Lemma 7.2.2, every failed node i must be accompanied by a working link $i-1 \to i+1$; hence the $(qp_2)^d$ term. Furthermore, by Theorem 7.2.3 (i), every pair of consecutive working nodes $(i, i+1)$, and there are $n-2d$ of them, must be accompanied by a working link $i \to i+1$; hence the term p_1^{n-2d}. Finally, in the case that all nodes are working and n odd, by Theorem 7.2.3 (ii), we have the additional case that all 2-links are working. But we have to subtract the probability of the case that all 1-links are working, which was already counted in the first term. ■

THEOREM 7.2.9 $\mathrm{CR}_L(0,1) = p_{01}$, $\mathrm{CR}_L(0,2) = p_1p_{01}p_{12} + q_1p_{12}$, $\mathrm{CR}_L(0,3) = p_1p_2p_{01}p_{23}(p_{12}+q_{12}p_{02}p_{13}) + p_1q_2p_{01}p_{13} + q_1p_2p_{02}p_{23}$, *and for* $n \geq 3$,

$$\begin{aligned}\mathrm{CR}_L(0,n+1) &= p_1p_{01}(q_{02}+p_{02}/p_{12})\mathrm{CR}_L(1,n+1) + \\ &\quad q_1p_{02}p_2\mathrm{CR}_L(2,n+1) - \\ &\quad (p_1p_{01}p_{02}q_2p_3p_{13}q_{12}/p_{12})\mathrm{CR}_L(3,n+1).\end{aligned}$$

Proof. The initial cases are easily verified. Consider the general case.

$$\begin{aligned}\mathrm{CR}_L(0,n+1) &= p_1p_{01}q_{02}\mathrm{CR}_L(1,n+1) + p_1p_{01}p_{02}q_2p_{13}p_3 \\ &\quad \mathrm{CR}_L(3,n+1) + p_1p_{01}p_{02}p_2\mathrm{CR}'_L(1,n+1) \\ &\quad + q_1p_{02}p_2\mathrm{CR}_L(2,n+1), \qquad (7.2.3)\\ \mathrm{CR}'_L(0,n+1) &= q_{02}\mathrm{CR}_L(1,n+1) + p_{02}q_2p_{13}p_3 \\ &\quad \mathrm{CR}_L(3,n+1) + p_{02}p_2CR'_L(1,n+1).\end{aligned}$$

Hence

$$\mathrm{CR}_L(0,n+1) = p_1p_{01}\mathrm{CR}'_L(0,n+1) + q_1p_{02}p_2\mathrm{CR}_L(2,n+1),$$

or

$$p_1p_{01}\mathrm{CR}'_L(0,n+1) = \mathrm{CR}_L(0,n+1) - q_1p_{02}p_2\mathrm{CR}_L(2,n+1). \qquad (7.2.4)$$

Replacing $(0,n+1)$ by $(1,n+1)$ and subtracting (7.2.4) into (7.2.3).

$$\begin{aligned}\mathrm{CR}_L(0,n+1) &= p_1p_{01}q_{02}\mathrm{CR}_L(1,n+1) + p_1p_{01}p_{02}q_2p_{13}p_3 \\ &\quad \mathrm{CR}_L(3,n+1) + (p_1p_{01}p_{02}/p_{12})[\mathrm{CR}_L(1,n+1) - \\ &\quad q_2p_{13}p_3\mathrm{CR}_L(3,n+1)] + q_1p_{01}p_2\mathrm{CR}_L(2,n+1) \\ &= p_1p_{01}(q_{02}+p_{02}/p_{12})\mathrm{CR}_L(1,n+1) + \\ &\quad q_1p_{02}p_2\mathrm{CR}_L(2,n+1) - \\ &\quad (p_1p_{01}p_{02}q_2p_{13}p_3q_{12}/p_{12})\mathrm{CR}_L(3,n+1).\end{aligned}$$

■

Thus $\mathrm{CR}_L(0, n+1)$ can be computed in linear time.

COROLLARY 7.2.10 *For IID,* $\mathrm{CR}_L(0) = p_1$, $\mathrm{CR}_L(1) = pp_1^2 + qp_2$, $\mathrm{CR}_L(2) = p^2p_1^2(p_1 + q_1p_2^2) + 2pqp_1p_2$, $\mathrm{CR}_L(n) = p(p_1q_2 + p_2)\mathrm{CR}_L(n-1) + qpp_2\mathrm{CR}_L(n-2) - p^2qp_2^2q_1\mathrm{CR}_L(n-3)$.

To compute $\mathrm{CR}_C(n)$, we first compute the reliability $CR_C^*(n)$ given that all nodes are working.

Index by $2i-2$ the link $i \to i+1$, and by $2i-1$ the link $i \to i+2$, $1 \le i \le n$. Consider the cyclic order of the $2n$ links. Then a node is an island if and only if two consecutive links are both failed. The probability that no island exists is simply the reliability $R_C(1, 2n)$ of a consecutive 2-out-of-$2n$ cycle. The probability of no two disconnected circuits is

$$\Pr(NDC) = (1-\epsilon_n)\left(\prod_{i=1}^{n} p_{i,i+2}\right)\left(\prod_{i \text{ odd}} q_{i,i+1} + \prod_{i \text{ even}} q_{i,i+1} - \prod_{i=1}^{n} q_{i,i+1}\right),$$

where $\epsilon = 1$ if n is odd and $= 0$ if n is even. Thus

$$\mathrm{CR}_C^*(1,n) = R(1,2n) - \Pr(NDC).$$

THEOREM 7.2.11 $\mathrm{CR}_C(1,n) = \left(\prod_{i=1}^{n} p_i\right)\mathrm{CR}_C^*(1,n) + q_1p_np_2p_{n2}\mathrm{CR}_L(2,n) + \sum_{m=2}^{n-1}\left(\prod_{i=1}^{m-1} q_mp_{m+1}p_{m-1,m+1}\right)\mathrm{CR}_L^*(1,m-1)\mathrm{CR}_L(m+1,n) + \left(\prod_{i=1}^{n-1} p_i\right)q_n p_{n-1,1}\mathrm{CR}_C^*(1,n-1)$.

Proof. If all nodes work $\mathrm{CR}_C(n) = \mathrm{CR}_C^*(n)$. Otherwise let m be the smallest index of a failed node. If $m = 1$, then nodes n and 2, link $n \to 2$ and the remaining $(n-1)$-line must all work. If $2 \le m \le n-1$, then node $m+1$, link $m-1 \to m+1$, the line from 1 to $m-1$ and the line from $m+1$ to 1 must all work. If $m = n$, then link $n-1 \to 1$ and the remaining $(n-1)$-line must work.

Note that CR_L^* can be computed from CR_L by setting $p_i = 1$ for all i.

■

It takes $O(n^2)$ time to compute $\mathrm{CR}_C(1,n)$ for the IND model and also for the IID model. The method of Theorem 7.2.11 also works for HR_C.

7.3 THE REVERSIBLE MODEL

Chen, Hwang and Li [57] studied the model in which a link can actually be traveled in both directions. While it can be easily argued that a path from source to sink would never traverse a 2-link backward, traversing a 1-link backward can help as shown in Figure 7.3.1 (only working nodes and links are shown):

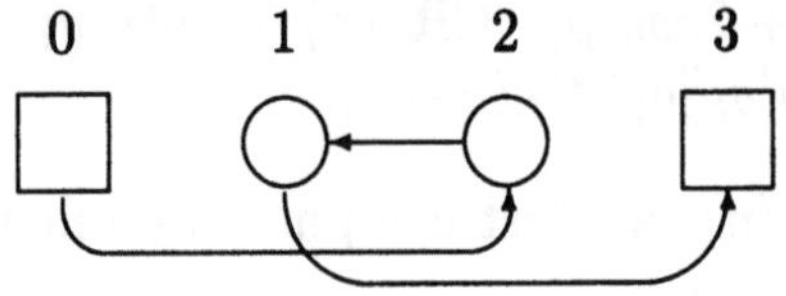

Figure 7.3.1. **A path with a backward link.**

In some situation, the forward link and the backward link between two nodes can be different entities and thus have independent states. For example, the availability of a flight from city A to city B doesn't imply the availability of the other direction. Even when the two links are derived from the same physical entity, like a pipeline, the two directions could have different reliabilities if the pipeline is located on a slope. Thus we assume $p_{i,i+1}$ and $p_{i+1,i}$ are not necessarily equal. In fact, we first assume independence of the forward state and the backward state, and then show that this independence assumption can be easily removed.

THEOREM 7.3.1 *For* $n \geq 4$,

$$\begin{aligned} R(0,1) &= p_{01}, \\ R(0,2) &= p_{02} + q_{02}p_1p_{01}p_{12}, \\ R(0,3) &= p_1p_{01}p_{13} + p_2p_{02}p_{23} - p_1p_{01}p_{13}p_2p_{02}p_{23} + \\ &\quad p_1p_2(q_{02}q_{13}p_{01}p_{23} + p_{02}p_{13}q_{01}q_{23}p_{21}), \\ R(0,4) &= p_1p_2[p_{01}(1-q_{02}q_{12}) + q_{01}p_{02}p_{21}][p_{24} + q_{24}p_3p_{34}(1-q_{13}q_{23})] \\ &\quad +p_1p_2p_3p_{01}p_{13}p_{32}q_{32}q_{12}(1-q_{24}q_{34}) + p_2p_{02}(q_1 + p_1q_{01}q_{21}) \\ &\quad (p_{24} + q_{24}p_3p_{23}p_{34}) + p_1p_{01}p_3p_{13}(q_2 + p_2q_{02}q_{12}q_{32})p_{34}, \end{aligned}$$

$$\begin{aligned} R(0,n+1) &= (1-q_{12}q_{32})^{-1}\{p_1[p_{02}(1-q_{01}q_{21}) + p_{01}q_{02}(1-q_{12}q_{32})] \\ &\quad R(1,n+1) + [p_2p_{02}(q_1 + p_1q_{01}q_{21})(1-q_{12}q_{32}) + \\ &\quad p_1p_{02}(1-q_{01}q_{21})p_2q_{13}q_{12}p_{32}]R(2,n+1) + \\ &\quad p_1p_{02}p_3p_{13}[p_{01}q_2(1-q_{12}q_{32}) - (1-q_{01}q_{21}) \\ &\quad (q_2 + p_2q_{12}q_{32})]R(3,n+1) + p_1p_2p_{02}(1-q_{01}q_{21}) \\ &\quad q_3p_{13}p_4p_{24}q_{12}p_{32}R(4,n+1)\}. \end{aligned}$$

Proof. $R(0,1), R(0,2)$ and $R(0,3)$, and $R'(0,1) = 1$, $R'(0,2) = 1 - q_{02}q_{12}$, $R'(0,3) = p_{13} + q_{13}p_2p_{23}(1-q_{02}q_{12})$ are easily verified.

Consider the four mutually exclusive and exhaustive events:

1. The source has a path to node 1 and a path to node 2 not through node 3.
2. The only path from the source to node 2 is $0 \to 1 \to 3 \to 2$.
3. The source has a path to node 2, but not to node 1.
4. The source has no path to node 2, but has a path $0 \to 1 \to 3$.

$$\begin{aligned}
R(0,n+1) &= p_1p_2[p_{01}(1-q_{02}q_{12})+q_{01}p_{02}p_{21}]R'(1,n+1)+ \\
&\quad p_1p_2p_3p_{01}p_{13}p_{32}q_{02}q_{12}R'(2,n+1)+ \\
&\quad p_2p_{02}(q_1+p_1q_{01}q_{21})R(2,n+1)+ \\
&\quad p_1p_{01}p_3p_{13}(q_2+p_2q_{02}q_{12}q_{32})R(3,n+1). \qquad (7.3.1)
\end{aligned}$$

$R(0,4)$ is verified by inserting $R'(1,4)$, $R'(2,4)$, $R(2,4)$ and $R(3,4)$ in (7.3.1). $R'(0,n+1)$ can be obtained from $R(0,n+1)$ by setting $p_1 = p_{01} = 1$. Thus

$$\begin{aligned}
R'(0,n+1) &= p_2(1-q_{02}q_{12})R'(1,n+1)+ \\
&\quad p_2p_3p_{13}p_{32}q_{02}q_{12}R'(2,n+1)+ \\
&\quad p_3p_{13}(q_2+p_2q_{02}q_{12}q_{32})R(3,n+1). \qquad (7.3.2)
\end{aligned}$$

Combining (7.3.1) and (7.3.2)

$$\begin{aligned}
R(0,n+1) &= p_1p_{01}R'(0,n+1)+p_1p_2q_{01}p_{02}p_{21}R'(1,n+1) \\
&\quad +p_2p_{02}(q_1+p_1q_{01}q_{21})R(2,n+1). \qquad (7.3.3)
\end{aligned}$$

Note that

$$p_2(1-q_{02}q_{12}) = p_2(1-q_{02}+q_{02}-q_{02}q_{12}) = p_2p_{02}+p_2q_{02}p_{12}$$

in (7.3.2). Using (7.3.3) to replace

$$p_2q_{02}[p_{12}R'(1,n+1)+p_3p_{13}p_{32}q_{12}R'(2,n+1)]$$

by

$$q_{02}[R(1,n+1)-p_3p_{13}(q_2+p_2q_{12}q_{32})R(3,n+1)],$$

(7.3.2) becomes

$$\begin{aligned}
R'(0,n+1) &= p_2p_{02}R'(1,n+1)+q_{02}[R(1,n+1)- \\
&\quad p_3p_{13}(q_2+p_2q_{12}q_{32})R(3,n+1)]+ \\
&\quad p_3p_{13}(q_2+p_2q_{02}q_{12}q_{32})R(3,n+1) \\
&= p_2p_{02}R'(1,n+1)+q_{02}R(1,n+1)+ \\
&\quad q_2p_{02}p_3p_{13}R(3,n+1). \qquad (7.3.4)
\end{aligned}$$

Solving $R'(0,n+1)$ and $R'(1,n+1)$ in (7.3.3) and (7.3.4), we obtain

$$\begin{aligned} R'(1,n+1) &= \{R(0,n+1) - p_1p_{01}q_{02}R(1,n+1) - p_2p_{02} \\ &\quad (q_1 + p_1q_{01}q_{21})R(2,n+1) - p_1p_{01}q_2p_{02}p_3 \\ &\quad p_{13}R(3,n+1)\}/\{p_1p_2p_{02}(1-q_{01}q_{21})\}. \qquad (7.3.5) \end{aligned}$$

$$\begin{aligned} R'(0,n+1) &= \{R(0,n+1) - p_1p_{01}q_{02}R(1,n+1) - \\ &\quad p_2p_{02}(q_1 + p_1q_{01}q_{21})R(2,n+1) - p_1p_{01}q_2 \\ &\quad p_{02}p_3p_{13}R(3,n+1)\}/\{p_1(1-q_{01}q_{21})\} \\ &\quad + q_{02}R(1,n+1) + q_2p_{02}p_3p_{13}R(3,n+1) \\ &= \{R(0,n+1) + p_1q_{02}q_{01}p_{21}R(1,n+1) - \\ &\quad p_2p_{02}(q_1 + p_1q_{01}q_{21})R(2,n+1) + p_1q_2p_{02} \\ &\quad p_3p_{13}q_{01}p_{21}R(3,n+1)\}/\{p_1(1-q_{01}q_{21})\} \end{aligned}$$

or

$$\begin{aligned} R'(1,n+1) &= \{R(1,n+1) + p_2q_{13}q_{12}p_{32}R(2,n+1) - \\ &\quad p_3p_{13}(q_2 + p_2q_{12}q_{32})R(3,n+1) + p_2q_3p_{13} \\ &\quad p_4p_{24}q_{12}p_{32}R(4,n+1)\}/\{p_2(1-q_{12}q_{32})\}. \qquad (7.3.6) \end{aligned}$$

Equating (7.3.5) with (7.3.6), we obtain

$$\begin{aligned} (1-q_{12}q_{32})R(0,n+1) &= p_1[p_{02}(1-q_{01}q_{21}) + p_{01}q_{02}(1-q_{12}q_{32})] \\ &\quad R(1,n+1) + [p_2p_{02}(q_1 + p_1q_{01}q_{21}) \\ &\quad (1-q_{12}q_{32}) + p_1p_{02}(1-q_{01}q_{21})p_2q_{13}q_{12}p_{32}] \\ &\quad R(2,n+1) + p_1p_{02}p_3p_{13}[p_{01}q_2(1-q_{12}q_{32}) - \\ &\quad (1-q_{01}q_{21})(q_2 + p_2q_{12}q_{32})]R(3,n+1) + p_1 \\ &\quad p_2p_{02}(1-q_{01}q_{21})q_3p_{13}p_4p_{24}q_{12}p_{32}R(4,n+1). \end{aligned}$$

■

COROLLARY 7.3.2 *For* $n \geq 4$,

$$\begin{aligned} R(n) &= p(1-q_1q_2)R(n-1) + pp_2[q + pq_1(1-p_1'p_2)]R(n-2) - \\ &\quad p^2q_1p_2^2(1-pp_1')R(n-3) + p^3qq_1p_1'p_2^3R(n-4). \end{aligned}$$

THEOREM 7.3.3 *Suppose* $\{p, p_1, p_1', p_2\} = \{n > s > t > u\}$. *Then* $p > p_2 > p_1 > p_1'$ *maximizes the reliability for n large.*

Proof. The characteristic polynomial is

$$\begin{aligned} F(x) &= x^4 - p(1-q_1q_2)x^3 - pp_2[pq_1(1-p_1'p_2) + q]x^2 + \\ &\quad p^2q_1p_2^2(1-pp_1')x - p^3qq_1p_1'p_2^3 \\ &= xf(x) + p^2q_1p_1'p_2^2g(x), \end{aligned}$$

where $g(x) = x^2 - px - pqp_2$. It can be shown that the largest root $\alpha >$ the largest root of $f(x) > p(1 - q_1q_2)$.

Suppose $0 \leq v' < v < 1$. Use F, f, g, α to denote the case that the assignment of v and v' is in the order of Theorem 7.3.3, and F', f', g', α' if not.

(i) $p_2 < p$. $F(x) - F'(x) = -(v-v')(x-vv')(p_1x^2 + v^2(v')^2q_1p_1')$ since $\alpha' > v'[1-(1-v)q_1] = v'[v+(1-v)p_1] \geq v'v$, $0 \geq F(\alpha') - F'(\alpha') = F(\alpha')$.

(ii) $p_1 < p_2$. $F(x) - F'(x) = (v - v')\{px(-x + p[1-(1-v)(1-v')]) + p^2p_1'(v + v' - vv')g'(x) - p^3qp_1'v^2(1-v')\}$.

It can be shown $g'(\alpha') \leq 0$. Further $\alpha' > p[1-(1-v)(1-v')]$. Hence $F(\alpha') - F'(\alpha) \leq 0$. ■

Next consider the dependency case. Suppose for link $i \to i+1$, the probabilities are $p_{fb}, p_f, p_b, p_\emptyset$ for working both ways, forward only, backward only and none. Observe that (i) A path never needs to contain a 1-link in both directions, (ii) all reliability functions we need are linear in $p_{i,i+1}$. When the forward and backward 1-link are the same physical entity and have the same state, we use p^* to denote its reliability and call it the 2-way model.

THEOREM 7.3.4 *The best permutation for the 2-way model is* $p > p_2 > p^*$.

Proof. (i) The proof of $p > p_2$ is as before.

(ii) $p_2 > p^*$. $F(x) - F'(x) = -(v - v')\{px[x - p(v + v' - vv')] + p^2vv'g(x) - p^3qv^2v'(1-v')\}$.

Since $\alpha' > p(v + v' - vv')$ and $g'(\alpha') \leq 0$, $F(\alpha) - F'(\alpha') \leq 0$. ■

7.4 THE $k \geq 3$ CASE

Both matrix methods introduced in Section 2.2 can be extended to the network model to compute reliabilities for general k systems. They work under any of the three definitions of network reliabilities, FR, CR and HR.

Hwang and Wright [117] extended to general k the transfer matrix method under CR for $k = 2$. Let $P_1(1, n)$ denote the probability that $G(s)$ contains no island and $P_2(1, n)$ that $G(s)$ contains no disconnected (strongly) circuits given that $G(s)$ contains no island. By Theorem 7.2.1,

$$R(1, n) = P_1(1, n) - P_2(1, n).$$

First consider the case that n is even. Partition the nodes and links into $n/2$ groups where group i consists of nodes $2i$, $2i - 1$ and their four outlinks, $1 \leq i \leq n/2$. Let s_i denote the state of group i under s, i.e., a specification of the states of all elements in group i. Then whether a node in group i is an island or not depends only on s_{i-1}, s_i and s_{i+1}.

Let $\mathcal{S}_i = \{s_{ij}\}$ denote the set of all possible states s_i such that s_i does not induce an island node in group i. For example, those states in which node $2i$ or $2i+1$ is working but both its outlinks are failed are not in $\mathcal{S}_i$. Furthermore, we can denote those states in which both nodes $2i$ and $2i+1$ are failed since node $2i-1$ must also fail (or be an island), and induction forces all nodes to fail. Thus $|\mathcal{S}_i| = 2^6 - 2^4 - 2^3 - 2^3 + 1 = 33$. Let $M_i = \{m_{xy}\}$ denote the transition matrix from $\mathcal{S}_i$ to $\mathcal{S}_{i+1}$ such that

$$m_{xy} = \begin{cases} 0, & \text{if } s_{ix} \cup s_{i+1,y} \text{ induces an island,} \\ \sqrt{\Pr(s_{ix})\Pr(s_{i+1,y})}, & \text{otherwise.} \end{cases}$$

The reason of taking the square root of the probability is that the state s_{ix} appears once in M_i (as rows) and once in M_{i-1} (as column), where $M_{n/2}$ is from $\mathcal{S}_{n/2}$ to $\mathcal{S}_1$. Then

$$P_1(1,n) = \text{Trace}\left(\prod_{i=1}^{n/2} M_i\right) \quad \text{(the line case needs some modification).}$$

Note that $P_1(n)$ can be computed in $O(|\mathcal{S}_i|^3 n)$ time for the IND model and $O(|\mathcal{S}_i|^3 \log n)$ time for the IID model, where $|\mathcal{S}_i|^3$ is a constant for n large.

If n is odd, then the last group has three nodes, six outlinks and consequently more states. But the overall complexity remains the same since only two matrices are affected.

The following lemma concerns the computation of $P_2(n)$.

LEMMA 7.4.1 *For n odd, $G(s)$ does not contain two disconnected circuits. For n even, two disconnected circuits exist if and only if one circuit consists of all odd nodes and the other all even nodes.*

Proof. Since each link covers a distance at most two, each circuit must contain at least $\lceil n/2 \rceil$ links. For n odd, these two disconnected circuits cannot coexist since $2\lceil n/2 \rceil > n$. For n even, each circuit contains $n/2$ b-links and the two circuits exhaust all nodes. ■

THEOREM 7.4.2

$$\text{CR}_C(1,n) = \begin{cases} \text{Trace}\left(\prod_{i=1}^{(n-1)/2} M_i\right), & n \text{ odd,} \\ \text{Trace}\left(\prod_{i=1}^{n/2} M_i\right) - \left(\prod_{i=0}^{n-1} p_i p_{i,i+2}\right) & \\ \left(\prod_{i=0}^{n-1} q_{2i,2i+1} + \prod_{i=0}^{n-1} q_{2i-1,2i} - \prod_{i=0}^{n-1} q_{i,i+1}\right), & n \text{ even.} \end{cases}$$

Proof. For n even, the second term gives the probability of two disconnected circuits: All nodes must work, either all 1-links from odd nodes fail, or all1-links from even nodes fail. ∎

The method of computing $P_1(1,n)$ can be readily extended to general k. The only difference is that the size of S_i grows with k. The computing of $P_2(1,n)$ needs some ad hoc analysis of enumerating disconnected circuits.

For the flow reliability, we have

THEOREM 7.4.3 $\mathrm{FR}_C(1,n) = P_1(1,n)$.

Proof. Since every working node has an outlink to another working node, a flow can always leave a node. But the number of nodes is finite. So, eventually, the flow must form a circuit. ∎

For the hamiltonian reliability, the extension to general k is straightforward for the line. The number of states in each S_i can be greatly reduced by removing any state in which link (i,j) does not exist for two adjacent working nodes i and j.

The condition that link (i,j) exists for every pair of adjacent working nodes is not necessary for the cycle since it is possible to hop around the cycle several times to collect all working nodes. For $k=2$ the only case this can happen is for odd n and the hops use all 2-links. For example, for $n=7$ the hamiltonian circuit is 1-3-5-7-2-4-6-1. Let $P_1'(n)$ denote the probability that link (i,j) exists for every pair of adjacent working nodes. Then

THEOREM 7.4.4 *For* $k=2$,

$$\mathrm{HR}_C(1,n) = P_1'(1,n) + \sum_{i=1}^{n} p_i p_{i,i+2}\left(1 - \prod_{i=1}^{n} p_{i,i+1}\right).$$

For $k \geq 3$, enumerating those wrapped-around circuits is much more difficult.

Next we discuss the Markov-chain method which works only for the line. S_i now is the set of working states for the k-window ending at node i. Chang, Chen and Hwang [46] observed that the number of states can be greatly reduced by not including the states of the links in the window state, but treating them as random variables represented in the transition probability matrix. This is legitimate since the reliability function is linear in each link.

For the flow reliability, a window-state is represented by the ordered set of nodes a flow reaches. The initial probability π (for the states in the first window) can be easily computed if k is not too large. The transition probability from $s \in S_i$ to $s' \in S_{i+1}$ is nonzero only if the states of nodes $i-k+3,\ldots,i$ are identical in s and s'. The only failure state is the state of empty set. More specifically, let a window state be denoted by a binary n-vector where a 1-entry at bit i represents that a flow can reach node i. Let M_i denote the transition matrix from S_i to S_{i+1}, where the empty state is listed first, and let $J = (0,1,\ldots,1)'$ be the $n \times 1$ column vector. Then

THEOREM 7.4.5 $FR_L(1,n) = \pi\left(\prod_{i=k}^{n} M_i\right)J.$

For the connectivity reliability, a window state is the ordered set of nodes which are not islands. For the hamiltonian reliability, a window state is the set of nodes which are not disconnected to either its previous or its succeeding working node. The rest of the two procedures are similar to the flow model.

Chapter 8

CONSECUTIVE-2 GRAPHS

So far we have talked about the linear system and the circular system for which the underlying configuration of the components is a line or a cycle. In this chapter we consider other configurations which can be best described by using the graph notion, although the links are not really there. Another way of looking at it is that the configuration is a graph with nodes and links, except that links cannot fail while nodes can. Although the concept of the graph system is valid for general k, so far, only the $k = 2$ case has been studied.

8.1 RELIABILITIES OF CONSECUTIVE-2 GRAPHS

A consecutive-2 graph works if and only if there does not exist two adjacent vertices both failed. Note that in this and the next section, only nodes can fail.

Let $R(G)$ denote the reliability of a consecutive-2 graph G. Santha and Zhang [222] proved

THEOREM 8.1.1 *Computing $R(G)$ is a #P-complete problem.*

Proof. A *vertex cover* is a set of vertices which covers every edge. Clearly G is working if and only if the set of working nodes is a vertex cover. Provan and Ball [211] proved that counting the number $c(G)$ of vertex covers of a graph G is #P-complete. Since $R(G) = 2^{-n}c(G)$ when $p = 0.5$, computing $R(G)$ is #P-complete even for $p = 0.5$. ∎

Santha and Zhang also gave an $O(n^2)$-time algorithm to compute the reliability of a tree. Du and Hwang [72] gave a linear-time algorithm for a wider class of graphs called *sequenceable graphs*. A graph is sequenceable if its vertices can be labeled by $\{1,2,\ldots,n\}$ satisfying the following conditions: Consider the subgraph G_i induced by the vertices $1,2,\ldots,i$, for any i. Let $C_{i1}, C_{i2},\ldots,C_{im_i}$ be the partition of G_i into m_i connected components. Suppose x is a vertex in C_{ij} adjacent to a vertex y not in G_i. Then every vertex $z > x$ in C_{ij} is also adjacent to y.

Note that a forest F is sequenceable. Let F consists of m trees $T_1,T_2,\ldots, T_m$ with $t_1,t_2,\ldots,t_m$ vertices. Set $s_i = \sum_{j=1}^{i} t_j$ for $1 \le i \le m$, and $s_0 = 0$. Assign labels from $s_{i-1}+1$ to s_i to the vertices of T_i. Represent each T_i as a rooted tree. Assign the largest label to the root and assign successively decreasing labels to other nodes following a depth first search. Figure 8.1.1 (a) illustrates the labeling of a tree. Another class of sequenceable graphs is defined by letting node i have links to the next k_i nodes, where each $k_i \le k_{i+1}+1$. An example is illustrated in Figure 8.1.1 (b).

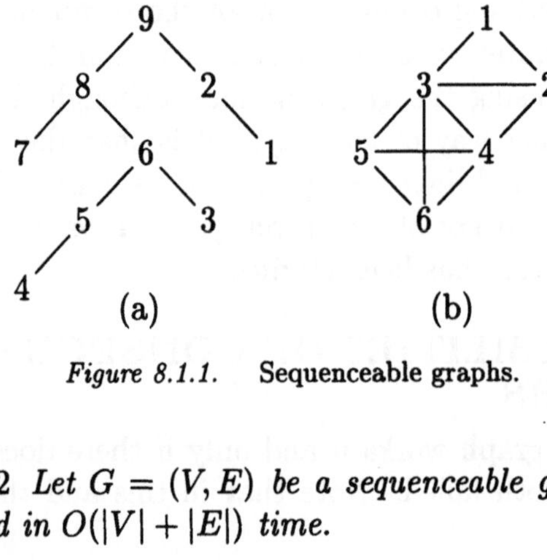

Figure 8.1.1. Sequenceable graphs.

THEOREM 8.1.2 *Let $G = (V,E)$ be a sequenceable graph. Then $R(G)$ can be computed in $O(|V|+|E|)$ time.*

Proof. A crucial observation is that if G is sequenceable, then G_i for all i and C_{ij} for all i and $j = 1,2,\ldots,m_i$ are sequenceable.

Theorem 8.1.2 is true for $|V| = n = 1$. The general n case is proved by induction. Let S_n denote the set of vertices adjacent to vertex n. Then

$$\begin{aligned} R(G_n) &= p_n R(G_{n-1}) + q_n\left(\prod_{k\in S_n} p_k\right)\left(\prod_{j=1}^{m_{n-1}} R(C_{n-1,j}\setminus S_n)\right) \\ &= p_n\left(\prod_{j=1}^{m_{n-1}} R(C_{n-1,j})\right) + q_n\left(\prod_{k\in S_n} p_k\right)\left(\prod_{j=1}^{m_{n-1}} R(C'_{n-1,j})\right), \end{aligned}$$

where $C'_{n-1,j}$ is obtained from $C_{n-1,j}$ by deleting those vertices in S_n. Let $C_{ij} = G(V_{ij}, E_{ij})$. By induction, $R(C_{ij})$ can be computed in $O(|V_{ij}|+$

$|E_{ij}|$) time. Furthermore, by the sequenceable property, C'_{ij} consists of vertices with the smallest labels in C_{ij}. Therefore, the computation of C'_{ij} is part of the recursive computation of $R(C_{ij})$. Since each C'_{ij} is associated with a distinct edge of G_n, the total number of multiplications in $\prod_{j=1}^{m_1}, \ldots, \prod_{j=1}^{m_{n-1}}$ is $|E|$. Finally, $\prod_{k \in S_1} p_k, \ldots, \prod_{k \in S_m} p_k$ can be computed in $O(n)$ time since there is a total of n p_k's. Therefore, $R(G_n)$ can be computed in linear time. ■

COROLLARY 8.1.3 *The reliability of a forest can be computed in linear time.*

There are graphs which are not sequenceable per se, but can be transformed to one after a few steps of computations. For example, C_n is not sequenceable, but

$$R(C_n) = p_n R(L(1, n-1)) + q_n p_1 p_{n-1} R(L(2, n-2)),$$

where $L(1, n-1)$ and $L(2, n-2)$ are sequenceable.

A (*two-terminal*) *series-parallel graph* is defined recursively by:
(i) An edge is a series-parallel graph.
(ii) A graph is series-parallel if it is a series or a parallel composition of two series-parallel graphs.

Figure 8.1.2 (a) shows a series-parallel graph, Figure 8.1.2 (b) a parallel composition and Figure 8.1.2 (c) a series composition.

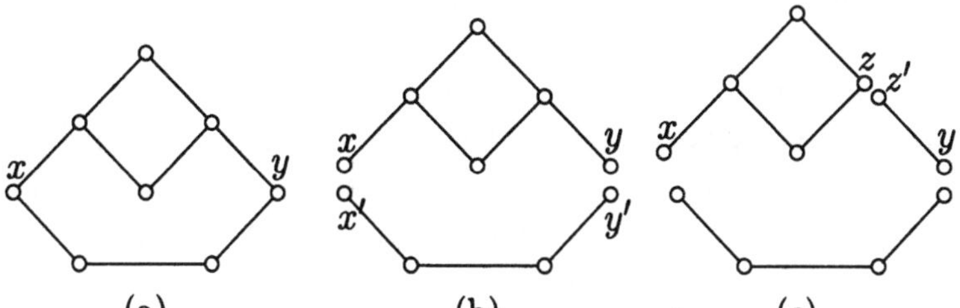

Figure 8.1.2. A series-parallel graph and compositions.

THEOREM 8.1.4 *Let $G = (V, E)$ be a series-parallel graph. Then $R(G)$ can be computed in linear time.*

Proof. To compute $R(G)$ for a series-parallel graph G with terminals x and y, we actually solve four similar problems with boundary conditions: for $i, j \in \{0, 1\}$, $R_{ij}(G)$ is defined to be the probability of G working under the conditions that x works (fails) if $i = 1$ $(i = 0)$ and y works (fails) if $j = 1$ $(j = 0)$. Note that

$$R(G) = q_x q_y R_{00}(G) + q_x p_y R_{01}(G) + p_x q_y R_{10}(G) + p_x p_y R_{11}(G).$$

For an edge xy, we have

$$R_{00}(xy) = 0 \text{ and } R_{01}(xy) = R_{10}(xy) = R_{11}(xy) = 1.$$

If G is obtained from a series composition of G_1 with terminals x and z and G_2 with terminal z and y, then

$$R_{ij}(G) = R_{i0}(G_1) q_z R_{0j}(G_2) + R_{i1}(G_1) p_z R_{1j}(G_2).$$

If G is obtained from a parallel composition of G_1 and G_2, then

$$R_{ij}(G) = R_{ij}(G_1) R_{ij}(G_2).$$

In either case, given $R_{i'j'}(G_1)$ and $R_{i''j''}(G_2)$, we can compute $R_{ij}(G)$ in a constant time. Each composition reduces the number of graph components by one. We start with $|E|$ graph components (links) and we end with one graph component G. Hence $O(|E|)$ time suffices. ■

8.2 OPTIMAL CONSECUTIVE-2 GRAPHS: GENERAL THEORY

Du and Hwang [71] established a basic theory of optimal consecutive-2 graphs. Consider $G = (V, E)$. Let θ denote an *automorphism*, i.e., a bijection from V to V such that $[\theta(u), \theta(v)] \in E$ if and only if $[u, v] \in E$. If $\theta^2 = 1$, then $\theta(u) = v$ is an equivalent relation and each equivalence class contains one or two vertices. Define a graph $G_\theta = (V_\theta, E_\theta)$, where V_θ is the set of equivalent classes $\{\tilde{u}\}$ of θ and $[\tilde{u}, \tilde{v}] \in E_\theta$ if and only if $[u, v] \in E$ and $[\theta(u), \theta(v)] \in E$. An automorphism θ is called *regular* if it satisfies the following three conditions:

(i) $\theta^2 = 1$.

(ii) G has no multiple edge, i.e., there do not exist four distinct edges $[u, v], [u, \theta(v)], [\theta(u), v]$ and $[\theta(u), \theta(v)]$ in E.

(iii) Let $(\tilde{u}_0, \tilde{u}_1, \ldots, \tilde{u}_m, \tilde{u}_{m+1} = \tilde{u}_0)$ denote a circuit in G_θ, and let u_i be an arbitrary element of $\tilde{u}_i$, $i = 0, 1, \ldots, m$. Then $|(\cup_{i=1}^{m} [u_i, u_{i+1}]) \cap E|$ is even.

Let P be an assignment of $\{p_1 < p_2 < \ldots < p_n\}$ to V. An edge $[\tilde{u}, \tilde{v}] \in E_\theta$ is called *singular* (or *nonsingular*) with respect to P if

$$[P(u) - P(\theta(u))][P(v) - P(\theta(v))] > 0 \text{ (or } \leq 0),$$

where $[u, v] \in E$. For example, if G is a line, then θ can be: $\theta(i) = j, \theta(j) = i, \theta(i+1) = j-1, \theta(j-1) = i+1, \theta(v) = v$ for all other $v \in V$. Edge $(i, i+1)$ is singular if $(P(i) - P(j))(P(i+1) - P(j-1)) > 0$.

THEOREM 8.2.1 *Let θ be a regular automorphism of G. Then for any assignment P there exists an assignment P_θ such that*
(i) $\{P_\theta(v), P_\theta(\theta(v))\} = \{P(v), P(\theta(v))\}$ *for all* $v \in V$.
(ii) *G_θ has no singular edge with respect to P_θ.*

Proof. If G_θ has no singular edge with respect to P, set $P_\theta = P$. Otherwise, let $[\tilde{u}_0, \tilde{v}_0]$ denote a singular edge of G_θ. Let E'_θ be the set of nonsingular edges. Then $\tilde{u}_0$ and $\tilde{v}_0$ are not connected in $G'_\theta = (V_\theta, E'_\theta)$. Suppose to the contrary that there exists a path $(\tilde{u}_0, \tilde{u}_1, \ldots, \tilde{u}_m = \tilde{v}_0)$. Let $u_i \in \tilde{u}_i$, $0 \leq i \leq m$, such that $[u_i, u_{i+1}] \in E$. Without loss of generality, assume $P(u_0) \geq P(\theta(u_0))$. Since $[\tilde{u}_i, \tilde{u}_{i+1}]$ are all nonsingular, $P(u_m) \geq P(\theta(u_m))$ for m even and $P(u_m) \leq P(\theta(u_m))$ for m odd. Since $[\tilde{u}_0, \tilde{u}_m]$ is singular,

$$[P(u_0) - P(\theta(u_0))][P(u'_m) - P(\theta(u'_m))] > 0,$$

where $u'_m \in \tilde{u}_m$ and $[u_0, u'_m] \in E$. So $u'_m = u_m$ if and only if m is even. In any case,

$$|\cup_{i=0}^{m} [u_i, u_{i+1}] \in E| = \begin{cases} m+1, & \text{for even } m, \\ m, & \text{for odd } m, \end{cases}$$

contradicting condition (iii) of a regular automorphism. Therefore $\tilde{u}_0$ and $\tilde{v}_0$ are not connected.

Let C be the component containing u_0. Define

$$P_1(u) = \begin{cases} P(u), & \text{if } u \in C, \\ P(\theta(u)) & \text{if } u \notin C. \end{cases}$$

Then all edges in $G'_\theta = (V_\theta, E'_\theta)$ are preserved. But a new edge $[u_0, v_0]$ is created. Repeating this process, eventually, we eliminate all singular edges. ∎

Let $U \subseteq V_\theta$ such that $\tilde{u} \in U$ implies $\tilde{u} = \{u, u'\}$ and $u \neq u'$. Let $S(U)$ denote a state of U such that exactly one component in each $\tilde{u} \in U$ is working and let $\overline{S}(U)$ denote its inverse. Let $S_\ell(U)$ ($S_s(U)$) denote the state that the component with the larger (smaller) reliability in each $\tilde{u} \in U$ is working.

THEOREM 8.2.2 *Suppose θ is a regular automorphism and P_θ is an assignment with no singular edge. Then*

(i) $P_\theta(G, S_\ell(U)) \geq P(G, S(U))$.
(ii) $P_\theta(G, S_\ell(U)) + P_\theta(G, S_s(U)) \geq P(G, S(U)) + P(G, \overline{S}(U))$.

Proof. True for $|V_\theta| = 1$. The general case is proved by induction on $|V_\theta|$.

Case 1. $U \neq \emptyset$. Let $\tilde{v} = \{v, v'\} \in U$. Let $N(\tilde{v})$ be the set of neighbors of $\tilde{v}$. For $\widetilde{W} \subseteq V_\theta$, define $W = \{w : w \in \tilde{w} \in \widetilde{W}\}$. Also define $G' = G \setminus (\{v, v'\} \cup W)$ and $U' = (U \setminus \{\tilde{v}\}) \cup (N(\tilde{v}) \setminus \widetilde{W})$.

$$
\begin{aligned}
\text{(i)} \quad P_\theta(G, S_\ell(U)) &= P_\theta(S_\ell(\tilde{v})) \sum_{\widetilde{W} \subseteq N(\tilde{v})} \Big(\prod_{w \in W} P_\theta(w)\Big) P_\theta(G', S_\ell(U')) \\
&\geq P(S(\tilde{v})) \sum_{\widetilde{W} \subseteq N(\tilde{v})} \Big(\prod_{w \in W} P(w)\Big) P(G', S(U')) \\
&= P(G, S(U)).
\end{aligned}
$$

$$
\begin{aligned}
\text{(ii)} \quad & P_\theta(G, S_\ell(U)) + P_\theta(G, S_s(U)) \\
= \ & P_\theta(S_\ell(\tilde{v})) \sum_{\widetilde{W} \subseteq N(\tilde{v})} \Big(\prod_{w \in W} P_\theta(w)\Big) P_\theta(G', S_\ell(U')) + \\
& P_\theta(S_s(\tilde{v})) \sum_{\widetilde{W} \subseteq N(\tilde{v})} \Big(\prod_{w \in W} P_\theta(w)\Big) P_\theta(G', S_s(U')) \\
= \ & [P_\theta(S_\ell(\tilde{v})) - P_\theta(S'_s(\tilde{v}))] \sum_{\widetilde{W} \subseteq N(\tilde{v})} \Big(\prod_{w \in W} P_\theta(w)\Big) P_\theta(G', S_\ell(U')) + \\
& P_\theta(S_s(\tilde{v})) \sum_{\widetilde{W} \subseteq N(\tilde{v})} \Big(\prod_{w \in W} P_\theta(w)\Big) [P_\theta(G', S_\ell(U')) + P_\theta(G', S_s(U')] \\
\geq \ & [P(S(\tilde{v})) - P(\overline{S}(\tilde{v}))] \sum_{\widetilde{W} \subseteq N(\tilde{v})} \Big(\prod_{w \in W} P(w)\Big) P(G', S(U')) + \\
& P(\overline{S}(\tilde{v})) \sum_{\widetilde{W} \subseteq N(\tilde{v})} \Big(\prod_{w \in W} P(w)\Big) [P(G', S(U')) + P(G', \overline{S}(U))] \\
= \ & P(G, S(U)) + P(G, \overline{S}(U)) \quad \text{by induction hypothesis (i), (ii).}
\end{aligned}
$$

Case 2. $U = \emptyset$. (both (i) and (ii) are reduced to $P_\theta(G) \geq P(G)$). Choose any $\tilde{v} = \{v, v'\} \in V_\theta$. Let $N'(\tilde{v}) = \{v : v \in \tilde{v} \in N(v)\}$, $G' = G \setminus \{v, v'\}$ and $G'' = G' \setminus W$. Define $\beta(\tilde{v}) = 0$ if $[v, v'] \in E$, $= 1$ otherwise.

$$
P_\theta(G) = P_\theta(v) P_\theta(v') P_\theta(G') + (1 - P_\theta(v))(1 - P_\theta(v'))\beta(\tilde{v})
$$

$$\left(\prod_{u \in N'(\tilde{v})} P_\theta(u))P_\theta(G'') + P_\theta(G, S_\ell(\tilde{v})) + P_\theta(G, S_s(\tilde{v})\right)$$
$$\geq P(v)P(v')P(G') + (1 - P(v))(1 - P(v'))\beta(\tilde{v})$$
$$\left(\prod_{u \in N'(\tilde{v})} P(u))P(G'') + P(G, S(\tilde{v})) + P(G, S(\tilde{v})\right)$$
$$= P(G) \qquad \text{by induction hypothesis (i) and (ii).}$$

■

COROLLARY 8.2.3 *The inequality in Theorem 8.2.2 is strict if* $P_\theta \neq P$ *and* $p_{[1]} > 0$.

To facilitate comparisons, sometimes we have to consider a supergraph $G' = (V', E')$ of $G = (V, E)$. Reliability are assigned to edges in $V' \setminus V$ by the following rules:

(i) $[u, v] \in E', u \in V, v \in V' \setminus V \Rightarrow p(v) = 1$.

(ii) $[u, v] \in E', u, v \in V' \setminus V \Rightarrow \{p(u), p(v)\} = \{1, p_1/2\}$.

Note that $R(G') = R(G)$.

Suppose a new node u is added at node $v \in V$ (v is called a *boundary node*). Since $p_u = 1$, the nonsingularity requires p_v to be smaller than its counterpart in any comparison. Moreover, all nodes having the same parity as v must be smaller, and all nodes having the other parity must be larger in their respective comparison. Therefore, if G has two boundary nodes, then the two nodes must have the same parity, otherwise they will force inconsistent requirements of being smaller or larger at a node. On the other hand, if we are comparing two graphs G and H and add new nodes at $v \in G$ and $u \in H$. Then u and v must have different parities (since $G' = H'$, we can unify the parities of the two graphs). All supergraphs we consider must meet these parity requirements.

For $u, v \in V$, we say that u, v are *comparable* if there exists a regular automorphism θ of G' feasible to G such that $\theta(u) = v$. If there exists a path x to u in G'_θ such that the pair in $\tilde{x}$ is already ordered, for example $x' \in V$ and $x \in V' \setminus V$, then order the pair in $\tilde{u}$ according to the parity of the number of edges from x to u to avoid singular edges. If not, we can order the pair in $\tilde{x}$ arbitrarily.

THEOREM 8.2.4 *If all pairs* u, v *are comparable, then* G *has an invariant assignment.*

Proof. To compare u and v, G' must be chosen such that u and v' are symmetric in G'. The ordering of u and v is then determined by the parities of the respective distances from u and v to the boundary nodes.

Note that the ordering is independent of the choice of the boundary nodes from the parity requirement in G'.

Therefore, if u and v are comparable, their ordering is fixed (in the case the ordering is arbitrarily set, both choices are equivalent in the sense that the system reliability is same due to symmetry). Furthermore, the orderings of the pairs are consistent since inconsistency implies the existence of two supergraphs violating the parity requirement. It follows that there exists a unique (up to equivalence) assignment with no singular index. By Theorem 8.2.2, it is invariant. ■

8.3 INVARIANT d-NARY TREES

We first study the system consisting of an m-line M and an n-line N. As a 1-line always takes the worst component, assume $m \geq n > 1$. If n is even, add $m-n$ nodes to either side. If n is odd, add any combination of $m-n$ nodes to both sides. Du and Hwang [69] proved the following results:

LEMMA 8.3.1 *Suppose $m > n$. For n even, the M line has the two largest (one if $m = 3$) and the two smallest reliabilities. For n odd, $N_i < \{M_i, \ldots, M_{m+1-i}\}$ for i odd and $N_i > \{M_i, \ldots, M_{m+1-i}\}$ for i even.*

COROLLARY 8.3.2 *$\{m, n\}$ is invariant if $3 \in \{m, n\}$. The 3-line is $(1, m+n, 2)$.*

Proof. If $m > n = 3$, $\{N_1, N_3\} < \{M_1, M_m\}, N_2 > \{M_2, M_{m-1}\}$.

If $m = n$, we may assume without loss of generality that $N_1 < \{N_3, M_1, M_m\}$. Then we get the same comparison as the $m > n$ case.

The $m = 3$ and $n = 2$ case follows from Lemma 8.3.1. ■

THEOREM 8.3.3 *For $m = n$ even, $p_i \in M$ for $i \equiv 0, 1 \pmod 4$ is invariant.*

Proof. Without loss of generality, assume $M_1 < \{M_m, N_1, N_n\}$. Consider the two mapping $M_i \to N_i$ and $M_i \to N_{n+1-i}$. We obtain

$$M_1 < N_1,\ N_n < M_m,\ M_2 > N_2,\ N_{n-1} > M_{m-1},$$

$$M_3 < N_3,\ N_{n-2} < M_{m-2},\ \ldots$$

Without loss of generality, assume $N_1 < N_n$. Then we have

$$M_1 < N_1 < N_n < M_m < M_3 < N_3 < N_{n-2}$$
$$< M_{m-2} < \ldots < M_{m-1} < N_{n-1} < N_2 < M_2.$$

■

THEOREM 8.3.4 *For $m = n+2$ even, $p_i \in M$ for $i \equiv 1, 2 \pmod 4$ is invariant.*

Proof. Consider the four mappings:

$$M_i \to N_i,\ M_{i+1} \to N_i,\ M_i \to N_{n+1-i},\ M_{i+2} \to N_{n+1-i}.$$

We obtain

$$\{M_i, M_{m+1-i}\} < \{N_i, N_{n+1-i}\} < \{M_{i+2}, M_{m-1-i}\} \text{ for } i \text{ odd},$$

and the above inequalities are reversed for i even. By Corollary 3.1.6, the four involved M_j's take consecutive p_ℓ's among the p_ℓ's of M. Hence we can order the M_j and N_j from small to large as

$$\{M_1, M_m\} < \{N_1, N_n\} < \{M_3, M_{m-2}\} < \{N_3, N_{n-3}\} < \dots$$

$$\{N_4, N_{n-3}\} < \{M_4, M_{m-4}\} < \{N_2, N_{n-1}\} < \{M_2, M_{n-1}\}.$$

Theorem 8.3.4 follows immediately. ■

Summarizing Corollary 8.3.2, Theorem 8.3.3 and Theorem 8.3.4, we have

THEOREM 8.3.5 *Let S be a system of lines in size $= 1, 3, n, n+2$ for some even n. Then the invariant assignment is the following:*

(i) *Assign the smallest p_i to the 1-lines.*

(ii) *Repeatedly assign the currently two smallest p_i and the largest p_i to a 3-line.*

(iii) *Suppose that there are x n-lines and y $(n+2)$-lines. Arrange the remaining p_i in order. Assign the first $2y$ to the $(n+2)$-lines, the next $2x$ to the n-lines, the next $2y$ to the $(n+2)$-lines, and so on, ending with assigning the last $2y$ to the $(n+2)$-lines.*

(iv) *Order the p_i's assigned to the n-lines with consecutive indices. Then the i-th line consists of p_j with $j = i$ or $2x - 1 - i \pmod{2x}$, $0 \le i \le x-1$.*

(v) *Order the p_i's assigned to the $(n+2)$-lines with consecutive indices. Then the i-th line consists of p_j with $j = i$ or $2y - 1 - i \pmod{2y}$, $0 \le i \le y-1$.*

Example 8.1. Suppose S consists of k k-lines for $k = 1, 2, 3, 4$. Then an invariant assignment is (1), (2, 30, 3), (4, 29, 5), (6, 28, 7), (8, 27, 20, 15), (9, 26, 21, 14), (10, 25, 22, 13), (11, 24, 23, 12), (16, 19), (17, 18).

THEOREM 8.3.6 *$\{m,n\}$ is not invariant if $5 \in \{m,n\}$, except for $\{5,3\}$.*

Proof. From Lemma 8.3.1, the 5-line has the two smallest and the two largest reliabilities except for $(5,3)$. Let c be the center element of the 5-line and let $\{a,b\}$ be the two extreme elements of the other line. If $p_8 \to 1$, then we need only compare the 2-cutsets. Since there are two 2-cutsets involving c, but only one involving a or b, we have $p_c > \{p_a, p_b\}$ in an optimal assignment. But if the two best components have reliability 1, then $p_c < \{p_a, p_b\}$ in an optimal assignment. ■

Shen and Zuo [241] studied the many-line case for the consecutive-k G system, namely, the system works if there exist k consecutive working components on every line. Since invariant assignment for a consecutive-k-out-of-n: G line exists only for $k \leq n \leq 2k$ (Theorem 3.6.6), they only considered the case that the length of each line is at most $2k$. They proved:

THEOREM 8.3.7 *The optimal assignment of $2mk$ components $p_{[1]} \leq p_{[2]} \leq \cdots \leq p_{[2mk]}$ to m consecutive-k-out-of-$2k$:G line is given by the algorithm:*

Step 1. Set $i = 1$. Let L_i be the i-th line.

Step 2. Assign $p_{[2(i-1)m+1]}, p_{[2(i-1)m+2]}, \ldots, p_{[2(i-1)m+m]}$ to L_1, L_2, $\ldots$, L_m, respectively.

Step 3. Assign $p_{[(2i-1)m+1]}, p_{[(2i-1)m+2]}, \ldots, p_{[2im]}$ to L_m, L_{m-1}, $\ldots$, L_1, respectively.

Step 4. Increment i and repeat Steps 2 and 3 until all components are assigned.

THEOREM 8.3.8 *The optimal assignment of $n = 2\ell k + 2m(k-1)$ components $p_{[1]}, p_{[2]}, \cdots, p_{[n]}$ to ℓ consecutive-$(k-1)$-out-of-$2(k-1)$:G lines and m consecutive-k-out-of-$2k$:G lines is given by the algorithm:*

Step 1. Add m components with $p = 1$ and m components with $p = 0$ to the n components.

Step 2. Use the optimal assignment of $2m + n$ components to $\ell + m$ consecutive-k-out-of-$2k$:G systems.

Example 8.2. $(p_{[3]}, p_{[7]}, p_{[8]}, p_{[4]})$, $(p_{[1]}, p_{[5]}, p_{[9]}, p_{[10]}, p_{[6]}, p_{[2]})$, is the optimal assignment for a consecutive-2-out-of-4 and a consecutive-3-out-of-6:G lines.

Recently, Chang and Hwang [44] studied the general case. Let $S(k_1, \ldots, k_m)$ denote a system of m lines, where L_i is a consecutive-k-out-$2k$:G system, and the system works if every line works. They proved

THEOREM 8.3.9 *$S(k_1, \ldots, k_m)$ has no invariant assignment except for the two cases given in Theorems 8.3.7 and 8.3.8.*

They also extend Theorem 8.3.9 to the system $S(k_1(n_1), \ldots, k_m(n_m))$, where L_i is a consecutive-k_i-out-n_i:G system, $n_i \leq 2k_i$.

THEOREM 8.3.10 *$S(k_1(n_1), \ldots, k_m(n_m))$ is invariant if and only if $S(n_1 - k_1, \ldots, n_m - k_m)$ is.*

A *complete d-regular tree* is a rooted *d*-regular tree such that all leaves are of the same heigh. Santha and Zhang [222] proved

THEOREM 8.3.11 *A complete d-regular tree is invariant for $d \geq 2$.*

Proof. Let i, j be two nodes. Because of the regularity, we can always add nodes (or subtrees) to the leaves to obtain the supertree in which i and j are symmetrical. Then use Theorem 8.2.4. ■

Example 8.3. To compare f and c, we can construct the supertree at right (which also compares b and a, and e and d).

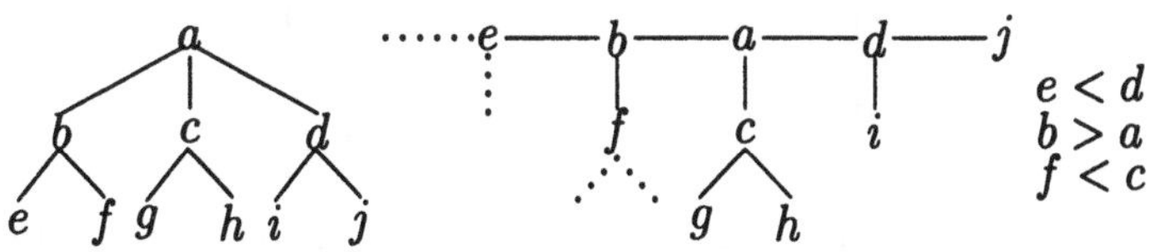

Figure 8.3.1. Example of a supertree.

A (d, m)-caterpillar consists of a path (called *body*) of m nodes each of degree d. Du and Hwang [71] stated without proof:

THEOREM 8.3.12 *A (d, m)-caterpillar is invariant.*

Proof. It suffices to consider the $d = 3$ case since all feet (of a given node) not on the body are symmetric.

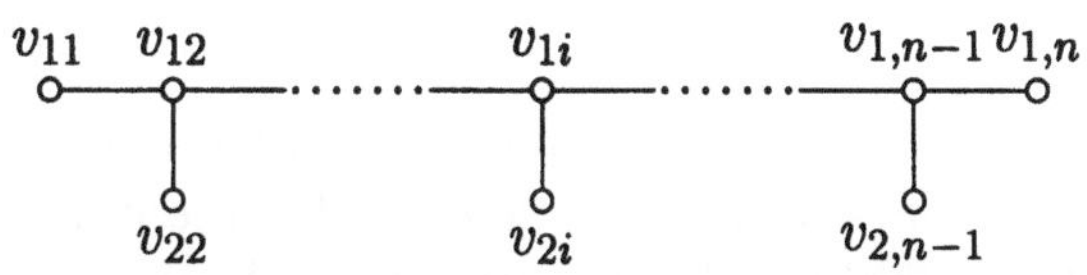

Figure 8.3.2. A standard caterpillar.

Represent a caterpillar in its standard form (see Figure 8.3.2) and label nodes on the top line (body) by $v_{11}, v_{12}, \ldots, v_{1n}$, and those on the bottom line (feet) by $v_{22}, v_{23} \ldots, v_{2,n-1}$. By symmetry, we may assume that $p_{11} = \min\{p_{11}, p_{1n}, p_{22}, p_{2,n-1}\}$. Then $p_{1i} < p_{1,n+1-i}$ for i odd, $p_{1i} > p_{1,n+1-i}$ for i even, $p_{2i} < p_{2,n-i}$ for i even, $p_{2i} > p_{2,n-1}$ for i odd. Furthermore, we can easily compare v_{1i} with v_{1j}, and v_{2i} with v_{2j} by adding on one side. It turns out the ordering of v_{1i} is same as the n-line, hence the ordering of v_{2i} is the reverse order. Therefore v_{13} is the smallest v_{1i} and v_{23} is the largest v_{2i}, $2 \le i \le n-1$. But the unique minimum cutset of v_{23}, (v_{13}, v_{23}) also contains v_{13}, which has other minimum cutsets like (v_{13}, v_{12}), (v_{13}, v_{14}). Hence $p_{13} > p_{23}$, which implies $p_{1i} > p_{2j}$ for all $2 \le i, j \le n-1$. We have established that $\{v_{2i}\}$ are the $n-1$ smaller reliabilities among $\{v_{2i}\} \cup \{v_{1i} : 2 \le i \le n-1\}$. Finally, $p_{12} > p_{1,n-1} \Rightarrow p_{22} < p_{2,n-1}$, and we may assume $p_{1n} < p_{2,n-1}$ by symmetry. Therefore $p_{1n} < p_{2i}$ for all $3 \le i \le n-1$. The invariant assignment is uniquely (up to symmetry) determined. ∎

Example 8.4.

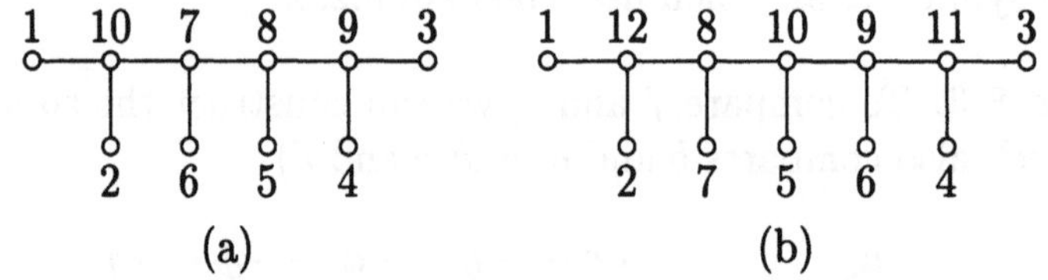

Figure 8.3.3. Invariant caterpillars.

Note that $v_{1,i+2}$ and v_{2i}, $3 \le i \le n-2$, are not comparable since any supergraph in which these two nodes are symmetric must include additions at v_{2i} and $v_{2,i+1}$. But $v_{2,i+1}$ and v_{2i} have different parities, violating the basic assumption of adding nodes. However, $v_{1,i+1}$ and $v_{2,i}$ can be ordered and the caterpillar is invariant. This example disproves a conjecture of Du and Hwang [71] that all nodes in an invariant consecutive-2 graph are comparable.

LEMMA 8.3.13 *The center of a star always gets the largest reliability.*

Proof. By criticality importance. ∎

THEOREM 8.3.14 *Let S consist of an m-star and an n-star, where $m \ge n$. Then the unique invariant assignment is that the center of the m-star gets the largest reliability and its m leaves get the m smallest.*

Proof. Add $m-n$ leaves to the n-star. The theorem follows immediately. ∎

Example 8.5.

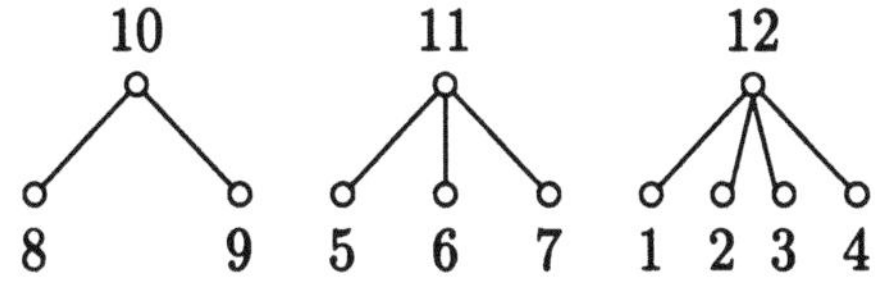

Figure 8.3.4. Invariant multi-star.

THEOREM 8.3.15 *Let S consist of an m-star and a (d, n)-caterpillar with $m \geq d$. Then the center of the m-star gets the largest reliability and its leaves the smallest.*

Proof. A star can be viewed as a 3-line when the middle node can have other children. Recall that mapping a 3-line with an n-line, the middle node of the 3-line always gets the largest. The existence of the children is consistent with this result since by adding children to node v of the n-line, p_v must be smaller than p of the middle node of the 3-line. ■

THEOREM 8.3.16 *If $S = (m$-line, n-line$)$ is invariant, then $S' = ((k, m)$-caterpillar, (k, n)-caterpillar$)$ is invariant.*

Proof. The order of nodes on the body determine those of the foot-nodes. ■

Example 8.6.

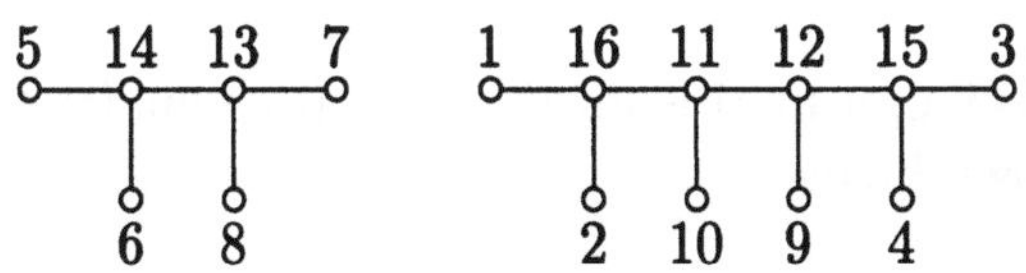

Figure 8.3.5. Invariant multi-caterpillar.

Call a complete d-regular tree of height h an (h, d)-tree.

THEOREM 8.3.17 *$S = \{(1, d)$-tree, (h, d)-tree$\}$, $h > 1$, is invariant.*

Proof. Represent the $(1, d)$-tree as shown in Figure 8.3.6 and map it to the last three levels of the (h, d)-tree. It is easily seen that $p_C < p_c, p_B > p_b$ and $p_A < p_a$. By Theorem 8.3.11, p_b is already the largest p_i, and p_a the smallest p_i in the (h, d)-tree. Hence p_B is the largest, and p_A the smallest, over both trees. By symmetry, the d leaves of the $(1, d)$-tree get the d smallest p_i. ■

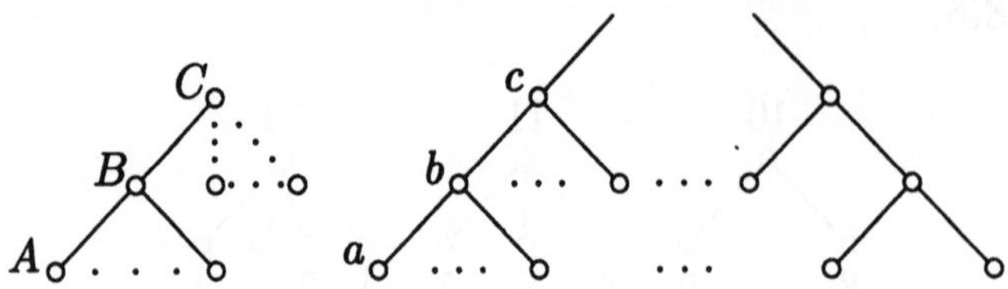

Figure 8.3.6. **Mapping the $(1, d)$-tree to the (h, d)-tree.**

For $h' > h \geq 2$, nodes are not comparable in general as shown below:

Example 8.7. To compare the root of the $(2, 3)$-tree with the leaves of the $(h, 3)$-tree, $h \geq 3$, we have to add new nodes at A' and a. But they have the same parity in the two supertrees, violating the basic assumption of adding nodes.

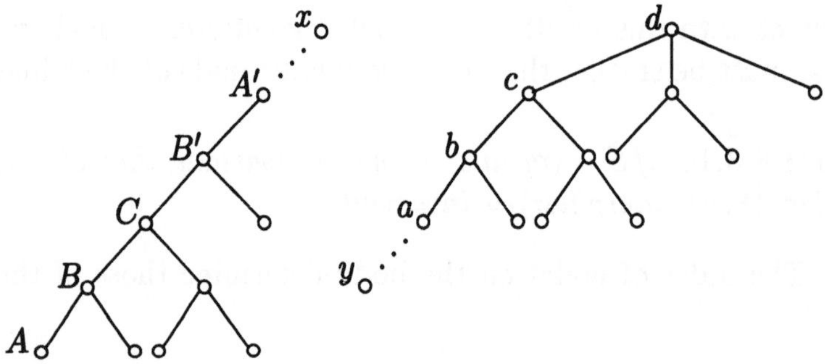

Figure 8.3.7. **Mapping the $(2, 3)$-tree to the $(h, 3)$-tree.**

Note that (m-line, n-line) is invariant $\not\Rightarrow$ (m-line, (n, d)-caterpillar) is invariant as shown below:

Example 8.8.

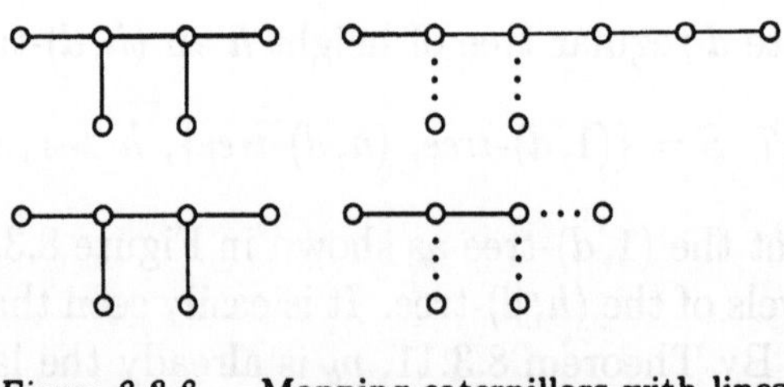

Figure 8.3.8. **Mapping caterpillars with lines.**

In both cases, the two nodes at which new nodes are added have different parities.

8.4 OTHER CONSECUTIVE-2 GRAPHS

The rooted d-nary tree, $d \geq 2$, is very much like a $(d+1)$-regular tree except that the root is of degree d. Let $T_t(d)$ denote a complete d-nary tree of height t. Santha and Zhang [222] proved that $T_t(2)$ is not invariant for $t \geq 3$ by using the notion of first-term optimal. An assignment is *first-term optimal* if it is optimal when $p_1 = p_n$ tends to 1. In other words, it is optimal when there are at most two failed nodes. We now extend their results to general d. Clearly, an optimal assignment must be first-term optimal. Let d_i denote the degree of node i.

LEMMA 8.4.1 *$d_i > d_j \Rightarrow p_i \geq p_j$ in the first-term optimal assignment.*

Proof. Under the first-term environment, the minimal cutsets dominate all other cutsets in probability. The set of minimal cutsets is the set of edges where each edge has the same probability of occurrence. Since node v contributes d_i cutsets, a larger reliability should be assigned to a node with a larger degree. ■

LEMMA 8.4.2 *The invariant assignment of $T_2(d)$ is that the d nodes at the middle level get the d largest p_i, and the root gets the $(d+1)$st largest.*

Proof. We only give the proof for $d = 2$.

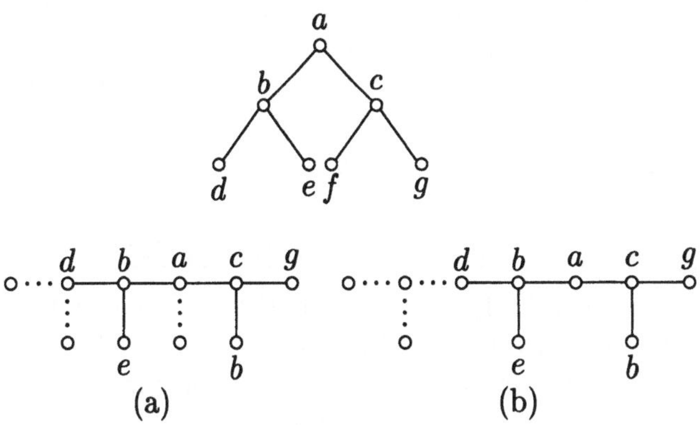

Figure 8.4.1. Comparisons of nodes in $T_2(d)$.

To compare a and b, consider Figure 8.4.1 (a). Then $b > a$ and $c > d$. To compare a and d, consider Figure 8.4.1 (b). Then $d < a$. We conclude Lemma 8.4.2 by symmetry of nodes. ■

THEOREM 8.4.3 *$T_t(d)$ is not invariant for $t \geq 3$.*

Proof. By Lemma 8.4.1, the level-t nodes should get the smallest p_i and the root the next smallest. Consider two cases that d^t p_i's are 0.

Then they are distributed to the level-t nodes. Clearly the level-$(t-1)$ nodes should get the largest p_i. Now consider the case that d^{t-1} p_i's are 1. From the above discussion, these p_i's should go to level-$(t-1)$ nodes, and the smallest d^t p_i's go to level-t nodes. The problem is then reduced to assigning the remaining p_i's to $T_{t-2}(d)$.

For $t = 3$, the root gets the larger p_i than the level-1 nodes. So does the $t = 4$ case by Lemma 8.4.2. On the other hand, by Lemma 8.4.1, the level-1 nodes should get larger p_i's than the root in $T_t(d)$ under the first-term environment. Hence $T_3(d)$ and $T_4(d)$ are not invariant.

We prove $T_t(d)$ is not invariant for $t \geq 5$ by induction. We have shown the existence of a set of p_i's such that $T_t(d)$ is reduced to $T_{t-2}(d)$. Furthermore, the reduction brings no restriction on the set of p_I's for $T_{t-2}(d)$. Therefore $T_{t-2}(d)$ is invariant implies $T_t(d)$ is invariant. ■

Not much is known outside of trees. The only two cases studied in the literature, cycles and cubes, are d-regular graphs. Du and Hwang [69] proved

THEOREM 8.4.4 *Consider a system of m cycles of sizes n, where n is even. An invariant assignment is repeatedly assigning the $n/2$ currently largest and $n/2$ currently smallest p_i's to one cycle.*

Proof. Consider $m = 2$. Let A and B be the two cycles. By the nonsingularity condition, $p_{A(1)} > p_{B(j)}$ forces $p_{A(1+\ell)} > p_{B(j+\ell)}$ for all even ℓ, and $p_{A(1+\ell)} < p_{B(j+\ell)}$ for all odd ℓ. Since this is true for all $j = 1, 2, \ldots, n$, Theorem 8.4.4 is true for $m = 2$. It is then easily verified that the assignment given in the theorem is the only one such that the assignment between every two cycles is invariant. ■

8.5 THE 2-DIMENSIONAL CASE

Salvia and Lasher [221] first studied the 2-dimensional consecutive system where nodes are arranged in an $n \times n$ grid. The grid system is considered failed if it contains an $k \times k$, $k < n$, square consisting of only failed components. Salvia and Lasher gave some simulation results as well as bounds for the reliability of such a system under the IID model.

Boehme, Kossou and Preuss [23] considered both linear and circular lattices (see Fig. 8.5.1), when the circular system can be obtained from the linear system by treating the last column and the first column as adjacent.

A lattice system fails if it consists of a connected configuration (prespecified) consisting of only failed components, called *bad*. The most studied configuration is an $r \times s$ rectangle. Let $R_L(r, s, m, n)$ denote the reliability of an $m \times n$ lattice not containing a bad $r \times s$ rectangle, and

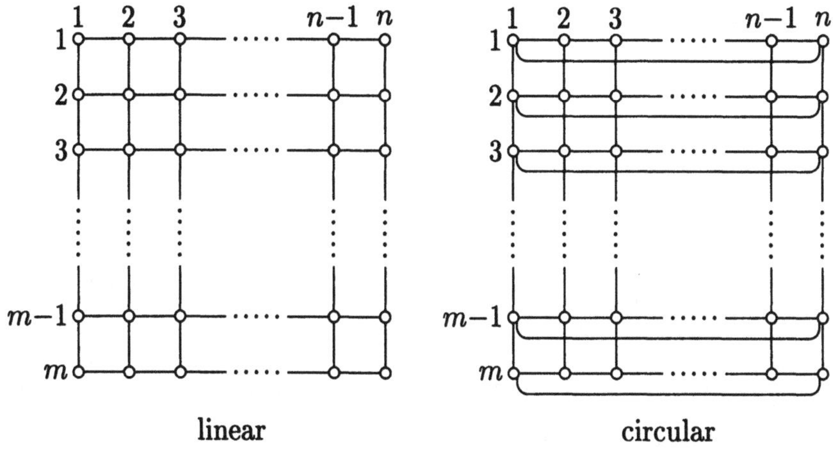

Figure 8.5.1. Linear and circular lattice systems.

let $R_C(r,s,m,n)$ denote the reliability of the circular counterpart. If the working probabilities of the components are important to be shown, we use $R_\Lambda(r,s,m,n;[p_{ij}])$ for an IND model and $R_\Lambda(r,s,m,n;[p])$ for an IID model.

THEOREM 8.5.1 *Consider the IND model. Define* $p_j = 1-\prod_{i=1}^{m}(1-p_{ij})$ *for* $1 \le j \le n$. *Then*

$$R_\Lambda(m,s,m,n;[p_{ij}]) = R_\Lambda(1,n;[q_j]),$$

where $\Lambda = L$ *or* C, *and* $R_\Lambda(1,n;[q_j])$ *is the reliability of a consecutive-s-out-of-n system.*

Proof. The m configurations in each column can be treated as one super-component which fails if and only if all m components fails. ∎

Yamamoto and Miyakawa [259] gave the following algorithm to compute $R_L(r,s,m,n)$. Define

$$\delta_{ij} = \begin{cases} 1, & \text{if components } (i,j),(i+1,j),\ldots,(i+r-1,j) \text{ all fail,} \\ 0, & \text{otherwise.} \end{cases}$$

Let $g = (g_1, g_2, \ldots, g_{m-r+1})$ and let $B(g_i,i,j)$ denote the event that the $r \times g_i$ rectangle whose upper right corner is (i,j) is bad. Let $R_L(r,s,m,n;g)$ denote the reliability of the (r,s,m,n) system given that $B(g_i,i,n)$ does not occur for each i. Then

$$R_L(r,s,m,n;g) = \sum_{x_n} R_L(r,s,m,n-1;g') \prod_{i=1}^{m} (p_{in}^{1-x_{in}} q_{in}^{x_{in}}), \quad (8.5.1)$$

where

$$g_i' = \begin{cases} g_i - 1, & \text{if } \delta_{in} = 1, \\ s, & \text{otherwise,} \end{cases}$$

and $x_n = (x_{1n}, x_{2n}, \ldots, x_{mn})$ is the 0-1 vector representing the state of the last column. Note that $R(r, s, m, n) = R(r, s, m, n; s\mathbf{1})$, where $\mathbf{1}$ is an $(m - n + 1) \times 1$ vector with all 1's.

We may partition $\sum_{x_n}$ in (8.5.1) into two parts, those without r consecutive failed components and those with. Then the first part can be written as $R_L(1, m; [p_{in}])R_L(r, s, m, n-1; s\mathbf{1})$.

There are s^{m-r+1} vectors of g, hence a table of ns^{m-r+1} different $R(r, s, m, n; g)$ to be computed. The first part needs computing n different $R(1, m; [p_{in'}])$ each in $O(m)$ time; hence a total of $O(nm)$ time. The second part has $2^m - N(r, m)$ cases, hence requiring $O(2^m)$ time. Combining the above, the total time required is $O(ns^{m-r+1}(nm + 2^m))$.

Noguchi, Sasaki, Yanagi and Yuge [180] gave the following recursive equation:

$$R_\Lambda(r, s, m, n) = R_\Lambda(r, s, m-1, n) - \Pr(\text{the last } r \text{ rows contains a bad } r \times s \text{ rectangle but not the first } m-1 \text{ rows}).$$

They computed the last term by enumeration.

Malinowski and Press [164] proved:

THEOREM 8.5.2 *The lower and upper bounds for the reliability of linear and circular (r, s, m, n) systems are*

$$\max\left(\prod_{i=1}^{m-r+1} \Pr\{A_\Lambda(i, r)\}, \prod_{j=1}^{n_\Lambda} \Pr\{B_\Lambda(j, s)\}\right)$$
$$\leq R_\Lambda(r, s, m, n; [p_{ij}])$$
$$\leq \min\left(\prod_{i=1}^{\lfloor m/r \rfloor} \Pr\{A_\Lambda((i-1)r + 1, r)\}, \prod_{j=1}^{\lfloor n/s \rfloor} \Pr\{B_\Lambda((j-1)s + 1, s)\}\right),$$

where $n_L = n - s + 1$, $n_C = n$; $\Pr\{A_\Lambda(w, r)\} = R_\Lambda(1, n; [\rho_{w,j}])$ is the reliability of a Λ consecutive-s-out-of-n:F system with component reliabilities

$$\rho_{w,j} = 1 - \prod_{k=w}^{w+r-1} (1 - p_{k,j}), 1 \leq j \leq n;$$

and $\Pr\{B_\Lambda(w, s)\} = R_L(1, m; [\sigma_{i,w}])$ is the reliability of a linear consecutive-r-out-of-m:F system with component reliabilities

$$\sigma_{i,w} = 1 - \prod_{\ell=w}^{w+s-1} (1 - p_{i,\ell}), 1 \leq i \leq m,$$

where $w + s - 1$ is added module n for the circular case.

COROLLARY 8.5.3 *For IID model, we have*

$$\begin{aligned}
&\max[(R_\Lambda(n;[1-(1-p)^r]))^{m-r+1}, (R_\Lambda(m;1-(1-p)^s))^{n_\Lambda}] \\
&\leq R_\Lambda(r,s,m,n;[p]) \\
&\leq \min[(R_\Lambda(n;[1-(1-p)^r]))^{\lfloor m/r \rfloor}, (R_\Lambda(m;[1-(1-p)^s]))^{\lfloor n/s \rfloor}].
\end{aligned}$$

Here we present a simple proof which is slightly different from Malinowski and Preuss's proof [164].

Proof. We only prove Corollary 8.5.3 as the argument for Theorem 8.5.2 is similar. Let

$$X_{ij} = \begin{cases} 1, & \text{if component } (i,j) \text{ works;} \\ 0, & \text{otherwise.} \end{cases}$$

$$Y_\ell = \begin{cases} 1, & \text{if rectangle } (X_{ij}, i = \ell, \ldots, r+\ell-1, j = 1, \ldots, n \text{ is bad.} \\ 0, & \text{otherwise.} \end{cases}$$

Therefore, the connected (r,s,m,n) system is a coherent system which consists of $Y_1, \ldots, Y_{m+r-1}$. By the definition of Y_i, $Y_1, \ldots, Y_{m+r-1}$ are associated. It follows from the property of the associated coherent system that

$$\prod_{\ell=1}^{m+r-1} E(Y_\ell) \leq R_L((r,s),(m,n);p).$$

On the other hand,

$$E(Y_\ell) = R_L(s,n;1-(1-p)^r).$$

Therefore,

$$R_L(s,n;1-(1-p)^r)^{m+r-1} \leq R_L((r,s),(m,n);p).$$

By the symmetry for columns and rows, we have

$$R_L(r,m;1-(1-p)^s)^{n-s+1} \leq R_L((r,s),(m,n);p).$$

Thus

$$\max[R_L(s,n;1-(1-p)^r)^{m+r-1}, R_L(r,m;1-(1-p)^s)^{n-s+1}] \leq R_L((r,s),(m,n);p).$$

Godbole, Potter and Sklar [89] gave an upper bound of $R(k,k,n,n)$ for the IID model. Let B_i be the event that the i-th $k \times k$ square is bad. They used a Jensen-type inequality due to Alon and Spencer [6] to obtain

$$\begin{aligned} R(k,k,n,n) &= \Pr\left\{ \cap_{i=1}^{(n-k+1)^2} \overline{B}_i \right\} \\ &\leq \left(\cap_{i=1}^{(n-k+1)^2} \Pr\{\overline{B}_i\} \right) \exp\left(\frac{\Delta}{2(1-q^{k^2})} \right) \\ &= \left(1-q^{k^2}\right)^{(n-k+1)^2} \exp\left(\frac{\Delta}{2(1-q^{k^2})} \right), \end{aligned}$$

where

$$\Delta = \sum_{i \neq j} \Pr\{B_i \cap B_j\} \leq (n-k+1)^2 q^{2k^2} \left(\sum_{a,b=1}^{k-1} 4q^{-ab} + \sum_{c=1}^{k-1} 4q^{-ck} \right) \equiv \Delta_0.$$

Since $4q^{-ab}$ is maximized (for $1 \leq a, b \leq k; a+b \leq 2k-2$) at $a = b = k-1$, the following is a slightly worse (but simpler) bound

$$\begin{aligned} \Delta_0 &\leq 4(n-k+1)^2 q^{2k^2} ((k^2-k+1) q^{-(k-1)^2} + q^{-k(k+1)}) \\ &= 4(n-k+1)^2 q^{k^2+k} ((k^2-k+1) q^{k-1} + 1). \end{aligned}$$

And so,

$$\begin{aligned} &R(k,k,n,n) \\ &\leq (1-q^{k^2})^{(n-k+1)^2} \exp\left(\frac{2(n-k+1)^2 q^{k^2+k} ((k^2-k+1) q^{k-1} + 1)}{1-q^{k^2}} \right). \end{aligned}$$

They also showed that this bound can be extended to the d-dimensional case for the IND model.

Besides computing the reliability, there are other related problems to be dealt with. We only provide some results without proofs since they are straightforward extensions of the one-dimensional case.

Koutras, Papadopoulos and Papastavridis [144] proved:

THEOREM 8.5.4 *Let T_n be the time to failure of a sequence of (k,k,n,n) systems with IID model, where k is fixed. If $t_n, n = 1, 2, \ldots$, is a sequence of positive real numbers such that $\lim_{n\to\infty} n^2 q^{k^2}(t_n) = \mu$, then*

$$\lim_{n\to\infty} \Pr\{T_n \geq t_n\} = \exp\{-\mu\}.$$

THEOREM 8.5.5 *Let T_n be the time to failure of a sequence of (k,k,n,n) systems with IID model, where k is fixed, whose component failure probability is of the form*

$$q(t) = (\mu t)^\alpha + o(t^\alpha)$$

for some positive real number α, then

$$\lim_{n\to\infty} \Pr\{n^{2/(\alpha k^2)} T_n \le t\} = 1 - \exp\{-(\mu t)^{\alpha k^2}\}.$$

Yamamoto and Miyakawa [259] gave the following result:

THEOREM 8.5.6 *Let μ, λ, η be constants, $n = \mu m^{\eta-1}$ and $p^{(m)} = 1 - \lambda m^{-\eta/(rs)}$, then*

$$\lim_{m\to\infty} R_L(r, s, m, n; [q^{(m)}]) = \exp(-\mu\lambda^{rs}),$$

where $\mu > 0$, $\lambda > 0$, $\eta > s$ or $r/(r-1) > \eta > 1$.

Zuo [268] discussed invariant optimal system design problems. Suppose the (m, s, m, n) system has mn components with distinct reliabilities ordered in increasing order as

$$p_{[1]} < p_{[2]} < \cdots < p_{[mn]}.$$

THEOREM 8.5.7 *For a linear (m, s, m, n):F or G system, necessary conditions for its optimal design are:*

1. $p_{i_1,j_1} < p_{i_2,j_2}$, $i_1, i_2 = 1, 2, \ldots, m$, $1 \le j_1 < j_2 \le \min\{s, n-s+1\}$;
2. $p_{i_1,j_1} < p_{i_2,j_2}$, $i_1, i_2 = 1, 2, \ldots, m$, $\max\{s, n-s+1\} \le j_1 < j_2 \le n$;
3. *If $s < n < 2s$, the $(2s-n)m$ most reliable components should be placed in columns $n-s+1, n-s+2, \ldots, s$ in any order.*

THEOREM 8.5.8 *For a linear $(2, s, 2, 2s)$:G system, the optimal arrangement is*

$$\begin{array}{cccccccccccc} 1 & 5 & 9 & 13 & \ldots & 4s-3 & 4s-1 & \ldots & 15 & 11 & 7 & 3 \\ 2 & 6 & 10 & 14 & \ldots & 4s-2 & 4s & \ldots & 16 & 12 & 8 & 4. \end{array}$$

For the circular case, the optimal arrangement is

$$\begin{array}{ccccccccccccc} 1 & 5 & 9 & 13 & \ldots & 4s-3 & 4s-1 & \ldots & 15 & 11 & 7 & 3 & 1 \\ 2 & 6 & 10 & 14 & \ldots & 4s-2 & 4s & \ldots & 16 & 12 & 8 & 4 & 2. \end{array}$$

Chapter 9

SOME RELATED SYSTEMS

In this chapter we discuss some isolated topics which are important but do not fit into the other chapters. Some of the covered topics are under fast development and will no doubt have their own chapters in the future.

9.1 CONSECUTIVELY CONNECTED SYSTEMS

Shanthikumar [237] extended the (linear) consecutive-k system to the consecutively connected system (CCS) by removing the requirement that the reachability parameter k is constant for all nodes. He assumed that node i can reach the next k_i nodes and gave an $O(n^2)$ algorithm. Hwang and Yao [119] further extended the CCS system to the multistate CCS (MCCS) by allowing a node to have various working states which affect its reachability. Thus node i has probability p_{ij} to be in state s_{ij} which allows node i to reach only the next j nodes. We assume that a node never fails as a receiving station. Note that p_{ii} is the probability that node i reaches no other node, hence can be considered as its unreliability.

Let $f(i,j)$ denote the probability that the source can reach node i (hence all nodes before i), and that node j is the farthest node which can be directly reached by a node in $\{0,1,\dots,i\}$. Then

$$f(0,j) = p_{0j}, \qquad 0 < j \le k_0,$$

$$f(i,j) = f(i-1,j)\sum_{h=i}^{j} p_{ih} + p_{ij}\sum_{h=i}^{j-1} f(i-1,h)$$

for $1 \le i \le n$ and $i < j \le \max_{0\le u\le i}(u+k_u)$, while $R_L(1,n) = f(n,n+1)$.

Suppose $k_i \leq k$ for all i. Then for each i we need to compute $f(i,j)$ for at most k j's. Furthermore, in computing $f(i,j)$, we treat

$$\sum_{h=i}^{j} p_{ih} = \sum_{h=i}^{j-1} p_{ih} + p_{ij}$$

and

$$\sum_{h=i}^{j-1} f(i-1,h) = \sum_{h=i}^{j-2} f(i-1,h) + f(i-1,j-1).$$

Since the sum terms on the right hand sides of the two equalities appeared in the computation of $f(i,j-1)$, the recursive step of computing $f(i,j)$ takes constant time. Therefore

THEOREM 9.1.1 *The reliability $R_L(1,n)$ for the MCCS can be computed in $O(kn)$ time.*

Hwang and Yao [119] proposed the circular CCS and suggested the circular version of flow reliability as well as the connectivity reliability. Actually, these two kinds of reliabilities are the same for the CCS. This is because the only time that a system without islands is not connected is due to the existence of disconnected circuits. But disconnected circuits do not exist in CCS since if node A_i of circuit A lies between nodes B_j and B_{j+1} on circuit B, then B_j connects to B_{j+1} implies B_j connects to A_i.

Hwang and Yao [119] extended Shanthikumar's algorithm [237] for the linear CCS to obtain an $O(mn^2)$ time algorithm for the circular CCS, where m is the minimum size of intersection between two adjacent minimum cutsets (in their algorithm the minimum cutsets can be linearly ordered).

Shanthikumar [237] proved that the elements of a minimum cutset is always consecutive. Let $M = (i, i+1, \ldots, i+m-1)$ be a minimum cutset. Let j be the first element of M on a working circuit C. Note that this implies $i, i+1, \ldots, j-1$ cannot reach j. Suppose the contrary that $v \in \{i, i+1, \ldots, j-1\}$ can reach j. Let u be the first node outside of M but preceding j on C. Then u can reach v directly. Hence we can expand C to include v, contradicting the definition of j. Define $P_{ij} = \sum_{h=i}^{j} p_{ih}$.

$$R_C(1,n) = \sum_{j=i}^{i+d-1} \left(\prod_{h=i}^{j-1} P_{ih}\right)\left(\sum_{u=j-k}^{i-1} (1-P_{u,j-1})\left(\prod_{v=u+1}^{i-1} P_{v,j-1}\right) R_L(j,u)\right).$$

We have to compute $R_L(j, i-1)$ for d j's. But the computation of $R_L(j, i-1)$ also yields $R_L(j, u)$ for all $j - k \leq u < i - 1$. Therefore it takes $O(dkn)$ time to compute all the required $R_L(j, u)$. The P_{ih} for $i \leq h \leq j-1$ can be computed in $O(d^2)$ time. For fixed j and u, $P_{v,j-1}$ for $u < v \leq i-1$ can be computed in $O(k)$ time. There are k u's and d j's. So P can be computed in at most $O(dk^2)$. Finally, with R_L and P as constants, R_C can be computed in $O(dk)$ time. Thus the total time is $O(dkn)$.

A system has *consecutive minimal cutsets* if the minimal cutsets can be linearly ordered into $c_1, c_2, \ldots$ such that $x \in c_i \cap c_j$ and $i < j \Rightarrow x \in c_h$ for all $i < h < j$.

For CCS, let $V(j)$ denote the set of predecessors of j. Then

THEOREM 9.1.2 $(V(k_0+1), V(k_0+2), \ldots, V(n+1))$ *is the sequence of consecutive minimum cutsets except that any cutset not minimal should be removed.*

Proof. These V-sets are clearly minimal cutsets. To show that there is no other minimal cutset, assume to the contrary that Z is such a cutset. Let z denote the largest component in Z. Since Z is minimal, $Z \setminus z$ is not a cutset. Hence the source can reach one of the nodes, say v, in $V(z+1)$. Note that this fact is true regardless whether z is working or not since $v < z$. Therefore for Z to be a cutset, necessarily $V(z+1) \subseteq Z$. So either $Z = V(z+1)$ or Z is not minimal.

Next we prove the V-sequence is consecutive. Suppose $x \in V(i)$ and $V(j)$, with $i < h < j$. Then x can reach i and j directly. Hence x can reach h directly. ∎

Shanthikumar [237] gave the following algorithm to compute the reliability of any system S with consecutive minimal cutsets. Let $C = (c_1, c_2, \ldots, c_r)$ be the sequence of minimal cutsets.

Step 1. Set $c_{r+1} = \emptyset$.
For $k = 1, 2, \ldots, m$, $m = 1, 2, \ldots, r$, obtain $D^k_m = c_k \cap \bar{c}_{m+1}$.

Step 2. Set $\beta^k_m = \prod_{j \in D^m_k} q_j$.

Step 3. Set $Q_0 = 0, Q_m = \sum\limits_{\ell=1}^{m} (1 - Q_{\ell-1})\beta^\ell_m$ for $m = 1, 2, \ldots, r$.

Step 4. $R = 1 - Q_r$.

It takes $O(|C|^2)$ to compute R.

Example. $c_1 = (1,2)$, $c_2 = (1,3)$, $c_3 = (1,4)$, $c_4 = (4,5)$.

$\beta_1^1 = q_2,$
$\beta_2^1 = q_2,\ \beta_2^2 = q_3,$
$\beta_3^1 = q_1q_2,\ \beta_3^2 = q_1q_3,\ \beta_3^3 = q_1,$
$\beta_4^1 = q_1q_2,\ \beta_4^2 = q_1q_3,\ \beta_4^3 = q_1q_4,\ \beta_4^4 = q_4q_5,$
$Q_1 = q_2,$
$Q_2 = q_2 + p_2q_3 = 1 - p_2p_3,$
$Q_3 = q_1q_2 + p_2q_1q_3 + p_2p_3q_1 = q_1,$
$Q_4 = q_1q_2 + p_2q_1q_3 + p_2p_3q_1q_4 + p_1q_4q_5.$

Q_i is the probability that all nodes in S_i is failed, where S_i is a minimum set containing for each j one node in c_j but not in c_{r+1}.

(1, 2), (1, 3), (1, 4), (4, 5)

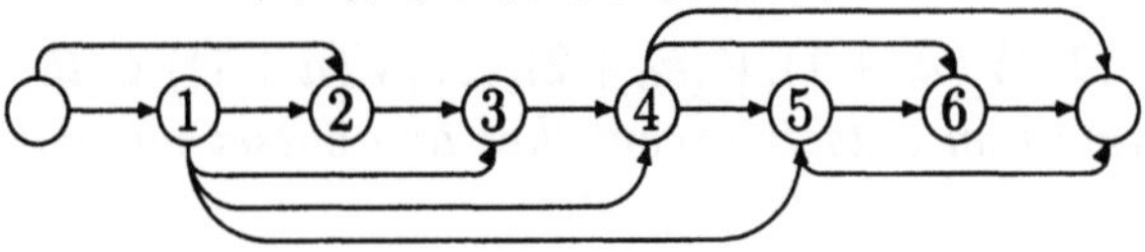

Hwang-Yao's circular MCCS algorithm.

Let $c_i \cap c_{i+1} = \{a_1, a_2, \ldots, a_m\}$. Let $t(x)$ denote the number of c_i containing x. Let $t(a_1) = \max t(a_i)$. Let C' be obtained from C by deleting a_1 from all c_i. Let C'' be obtained from C by deleting all cutsets containing a_1. Then

$$R(C) = (1 - p_{a_1})R(C') + p_{a_1}R(C'').$$

We prove that $C'' = (c_{i+x}, c_{i+x+1}, \ldots, c_r, c_1, \ldots, c_{i-y})$ is linear, i.e., $c_{i+x} \cap c_{i-y} = \emptyset$. Suppose to the contrary that $z \in c_{i+x} \cap c_{i-y}$. Then z is in $c_{i-y}, c_{i-y+1}, \ldots, c_{i+x}$, or $z \in \{a_1, a_2, \ldots, a_m\}$ and $t(z) \geq t(a_1) + 2$, a contradiction.

While $R(C')$ may still be circular, $|c_i \cap c_{i+1}|$ is reduced by one. So in m steps, $c_i \cap c_{i+1} = \emptyset$, and the sequence $(c_{i+1}, c_{i+2}, \ldots, c_r)$ is linear.

Given a set of minimal cutsets, we can reconstruct the system in $O(kn)$ time. Then we can compute R_L in $O(kn)$ time. Therefore the total time is $O(mkn)$.

Example. $C = (12, 23, 34, 45, 51)$, $c_1 \cap c_2 = 2$, $C' = (1, 3, 34, 45, 51)$, $C'' = (34, 45, 51)$.

Zuo and Liang [271] gave an $O(kdn)$-time algorithm for the circular MCCS where d is the minimum size of a cutset (they confused d with m).

Hwang and Yao [119] also proposed the 2-*way model* in which each node i has probability $p_i(\ell, r)$ of reaching ℓ preceding nodes and r succeeding nodes. For the linear system, the system works if source can

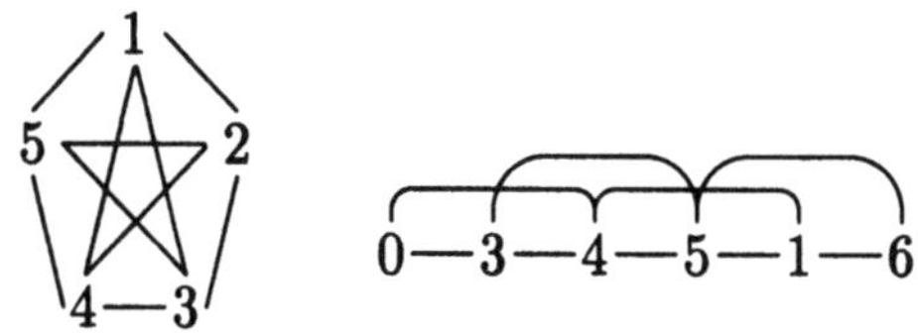

Figure 9.1.1. Reducing a circular set of cutsets to linear.

reach sink and vice versa. For the circular system, the system works if i can reach j in both directions.

9.2 MULTI-FAILURE CONSECUTIVE SYSTEMS

Koutras [141] considered two component-failure modes, open-circuit and short-circuit, as well as two system failure modes: consecutive k open-circuits and consecutive r short-circuits. Let q_i and q_i' denote the two failure probabilities of node i, with $p_i + q_i + q_i' = 1$. Let $R_1(n)$ and $R_2(n)$ denote the system reliability when the last component is open and short, respectively.

THEOREM 9.2.1 $R(1,n) = p_n R(1,n-1) + q_n R_1(1,n) + q_n' R_2(1,n)$,

$$\begin{aligned} R_1(1,n) &= R(1,n-1) - \left(\prod_{j=n-k+1}^{n-1} q_j\right) \\ &\quad [p_{n-k}R(1,n-k-1) + q_{n-k}' R_2(1,n-k)], \\ R_2(1,n) &= R(1,n-1) - \left(\prod_{j=n-r+1}^{n-1} q_j'\right) \\ &\quad [p_{n-r}R(1,n-r-1) + q_{n-r}' R_1(1,n-r)]. \end{aligned}$$

For the IID model, we can determine R_1 and R_2.

THEOREM 9.2.2 *Let δ_{ij} denote Kronecker's delta, i.e. $\delta_{ii} = 1$ and $\delta_{ij} = 0$ for $i \neq j$. Then $R(n) = R(n-1) - (p+q')q^k R(n-k-1) - (p+q)q'^r R(n-r-1) + q^k q'^r R(n-k-r) + pq^k q'^r R(n-k-r-1) + \delta_{n0} - q^k \delta_{nk} - q'^r \delta_{nr} + q^k q'^r \delta_{n,k+r}$.*

Proof. Note that

$$\begin{aligned} G(z) &= pz\left(G(z) + \frac{1}{z}\right) + qG_1(z) + q'G_2(z), \\ G_1(z) &= z\left(G(z) + \frac{1}{z}\right) - q^k\left[pz^{k+1}\left(G(z) + \frac{1}{z} + \ldots + \frac{1}{z^{k+1}}\right) + \right. \end{aligned}$$

$$
\begin{aligned}
& q'\left(G_2(z) + \frac{1}{z} + \ldots + \frac{1}{z^{k+1}}\right)\Bigg], \\
G_2(z) &= z\left(G(z) + \frac{1}{z}\right) - q'^r\left[pz^{r+1}\left(G(z) + \frac{1}{z} + \ldots + \frac{1}{z^{r+1}}\right) + \right. \\
& \left. q\left(G_1(z) + \frac{1}{z} + \ldots + \frac{1}{z^{r+1}}\right)\right].
\end{aligned}
$$

Solving the three variables $G(z), G_1(z), G_2(z)$ in the three equations, we obtain Theorem 9.2.2. ∎

Koutras [141] also gave upper and lower bounds.

Theorem 9.2.3 $\prod_{i=k}^{n}\left(1-\prod_{j=i-k+1}^{i} q_j\right)+\prod_{i=r}^{n}\left(1-\prod_{j=i-r+1}^{i} q'_j\right)-1 \leq R(1,n) \leq$ $\prod_{i=\min\{k,r\}}^{n}\left(1-p_{i-k}\prod_{j=i-k+1}^{i} q_j - p_{i-r}\prod_{j=i-r+1}^{i} q'_j\right).$

Proof.

$$
\begin{aligned}
1 - R(1,n) &\leq \text{Pr(system fails in open node)} + \\
& \quad \text{Pr(system fails in short node)} \\
&\leq 1-\prod_{i=k}^{n}\left(1-\prod_{j=i-k+1}^{i} q_j\right) + 1-\prod_{i=r}^{n}\left(1-\prod_{j=i-r+1}^{i} q_j\right).
\end{aligned}
$$

Hence the lower bound.

Define $R_1(1,n), R_2(1,n)$ as the reliability of the (n,k,d) system conditional on the nth component being the open-failure mode and the short-failure mode, respectively. Then

$$
\begin{aligned}
R_1(1,n) &= R(1,n-1) - \\
& \quad \left(p_{n-k}R(1,n-k-1) - q'_{n-k}R_2(1,n-k)\right)\prod_{j=n-k-1}^{n-1} q_j \\
&\leq R(1,n-1) - p_{n-k}R(1,n-k-1)\prod_{j=n-k-1}^{n-1} q_j \\
&\leq R(1,n-1)\left(1-p_{n-k}\prod_{j=n-k-1}^{n-1} q_j\right).
\end{aligned}
$$

Similarly,

$$
R_2(1,n) \leq R(1,n-1)\left(1-p_{n-r}\prod_{j=n-r-1}^{n-1} q_j\right).
$$

Hence

$$\begin{aligned} R(1,n) &= p_n R(1,n-1) + q_n R_1(1,n) + q_n' R_1(1,n) \\ &\le \Big(1 - p_{n-k} \prod_{n-k+1}^{n-1} q_j - p_{n-r} \prod_{n-r-1}^{n-1} q_j'\Big) R(1,n-1). \end{aligned}$$

The upper bound follows from the recursive inequality immediately. ■

Koutras indicated that this 2-failure model can be extended to multi-failure.

Sun and Liao [243] defined an (n, d, k) system to be an n-line or n-cycle which fails if either some d components fail or some k consecutive components fail. They computed the reliability for the IID model.

THEOREM 9.2.4 $R_x(n,d,k) = \sum_{j=0}^{d} N_x(d,n,k) p^{n-j} q^j$ *for* $x \in \{L, C\}$.

Chang, Cui and Hwang [38] did the IND case. If $d \le k$, then the (n, d, k) system is just the d out of n system. So assume $d > k$.

They gave a matrix algorithm. Define the state space for Markov chain $\{Y(t)\}$ as

$$S = \{(i,j) : 0 \le i \le k-1, i \le j \le d-1\} \cup \{s_N\},$$

where (i, j) indicates a working state in which the last i component fail and the system has j failed components, and s_N indicates the state in which the system fails. Note that $N = |S| = (2d - k + 1)k/2 + 1$.

For convenience, re-label state (i, j) as state $s_{(2d-i-1)i/2+j+1}$ for $0 \le i \le k-1, i \le j \le d-1$. Then the transition matrix of $\{Y(t)\}$ is

$$\Lambda_t = \begin{pmatrix} A^{(0)} & B^{(0)} & \mathbf{0} & \dots & \mathbf{0} & C^{(0)} \\ A^{(1)} & \mathbf{0} & B^{(1)} & \dots & \mathbf{0} & C^{(1)} \\ \vdots & \vdots & \vdots & \ddots & \vdots & \vdots \\ A^{(k-1)} & \mathbf{0} & \mathbf{0} & \dots & B^{(k-1)} & C^{(k-1)} \\ \mathbf{0} & \mathbf{0} & \mathbf{0} & \dots & \mathbf{0} & 1 \end{pmatrix}_{N\times N},$$

where

$$A^{(i)} = \begin{pmatrix} 0 & \dots & 0 & p_t & \dots & 0 \\ \vdots & \dots & \vdots & \vdots & \ddots & \vdots \\ 0 & \dots & 0 & 0 & \dots & p_t \end{pmatrix}_{(d-i)\times d} \qquad \text{for } 0 \le i \le k-1,$$

$$B^{(i)} = \begin{pmatrix} p_t & \dots & 0 \\ \vdots & \ddots & \vdots \\ 0 & \dots & p_t \\ 0 & \dots & 0 \end{pmatrix}_{(d-i)\times(d-i-1)} \qquad \text{for } 0 \le i \le k-2,$$

$$B^{(k-1)} = \mathbf{0}_{(d-k+1)\times(d-k)},$$

$$C^{(i)} = \begin{pmatrix} 0 \\ \vdots \\ 0 \\ q_t \end{pmatrix}_{(d-i)\times 1} \quad \text{for } 0 \le i \le k-2 \text{ and } C^{(k-1)} = \begin{pmatrix} q_t \\ \vdots \\ q_t \end{pmatrix}_{(d-k+1)\times 1}.$$

Let $\pi_0 = (1, 0, \ldots, 0)$ denote the initial probabilities. Then the reliability $R_L = \pi_0 \prod_{t=1}^{n} \Lambda_t(n) U_0^{\mathrm{T}}$, where $U_0 = (1, \ldots, 1, 0)$. It takes $O(nN^2)$ time to compute R_L. For the circular case, an analogous argument leads to

$$R_C = \sum_{i=0}^{k-1} p_{n-i} \sum_{m=0}^{i-1} q_{n-m} \pi_i \prod_{i=1}^{n-i-1} \Lambda_t(n-i-1) U_0^{\mathrm{T}},$$

where π_i is an N-vector with 1 at position $id - i(i+1)/2 + 1$, and 0 elsewhere. R_C can be computed in $O(knN^2)$ time.

Recently, Chang, Chen and Hwang [47] gave a better implementation of the matrix algorithm. By using a data structure which keeps only the nonempty entries in each column along with their (row) positions, and by noting that each Λ_t has only $2N - 1$ nonempty entries, we can multiply an N-vector with Λ_t with only $2N-1$ multiplications. Namely, if column C has only ℓ nonempty entries with positions $r_1, \ldots, r_\ell$, then we multiply these entries with their corresponding bits, bits $r_1, \ldots, r_\ell$, in the N-vector. This way it takes $O(N)$ time to compute π_1 from $\pi_0\Lambda_1$, π_2 from $\pi_1\Lambda_2$ and so on, with a total of $O(nN)$ time to compute R_L and $O(knN)$ time to compute R_C. A critical observation is that each time we must multiply a vector with a Λ_t to take advantage of the sparseness of Λ_t. If we multiply Λ_t among themselves first, then the sparseness is lost. Since $d \le 2d - k \le 2d$, $O(N) = O(kd)$.

Chang, Chen and Hwang [47] also gave the following equations whose justifications are similar to those given in Section 2.1 for the classical model. For fixed k, let $R(d; 1, n)$ denote the reliability of an (n, d, k) system.

THEOREM 9.2.5 $R_L(d; 1, n) = \sum_{i=n-k+1}^{n} p_i \Big(\prod_{j=i+1}^{n} q_j \Big) R_L(d-n+i; 1, i-1).$

$$R_C(d; 1, n) = \sum_{1 \le i \le k} \sum_{n-k-i \le j \le n} p_j \Big(\prod_{h=j+1}^{n+i-1} q_h \Big) p_i R_L(d-(n+i-j-1); i+1, j-1).$$

Using the same argument as in Section 2.1, it can be verified that $R_L(d; 1, n)$ can be computed in $O(kdn)$ time and $R_C(d; 1, n)$ in $O(k^2dn)$ time.

Finally, by using automata theory, Chang, Chen and Hwang showed that the number of states can be reduced from N to $N' = (d-k+1)k+1$. Thus the complexities become $O(nN')$ and $O(knN')$ for the linear and circular case, respectively. Note that for $d-k$ a constant, $O(N') = O(k)$, thus saving a factor of d from $O(N)$.

9.3 REDUNDANT CONSECUTIVE-k SYSTEMS

In a redundant consecutive-k system, r_i components are placed at node i which fails if and only if all r_i components fail. Let $S(k,n,\{r_i\})$ denote such a system and $R_S(k,n,\{r_i\})$ its reliability. $S(k,n,\{f\})$ (or $S(k,n,f)$ for short) is also known as an f-fold system. Hwang and Shi [115] proved

THEOREM 9.3.1 *$R_S(k,n) \leq R_S(1,n,\{k\})$ for $S \in \{L,C\}$.*

Proof. $C_S(k,n) \subset C_S(1,n,\{k\})$. ∎

$R_S(1,n,\{r_i\})$ can be computed by the same methods for the consecutive-k system except setting $q_i = \prod\limits_j q_{ij}$.

They also proved

THEOREM 9.3.2 *$S(k,n,f)$ is not invariant for $2 \leq k \leq n-1$ except for $S(n-1,n,2)$.*

Proof. The noninvariance of $S(k,n,f)$ clearly implies the noninvariance of $S(k,n,f')$ for $f < f'$. Since $S(k,n,1)$ is not invariant except for $k = 2, n-2$ and $n-1$, it suffices to consider only these three cases.

For $k = 2$, it is easy to construct example that the largest and the smallest reliability are assigned to the same node in an optimal assignment. Under invariance, that should always be the case. Let x be the node receiving the two extreme reliabilities and let y receiving reliabilities p and p'. Then in general we don't know whether the reliability r_x of x is larger than reliability r_y of y. But an invariant permutation of the nodes rely on knowing the ordering of the node reliabilities.

For $k = n-1$, the $2k$ smallest reliabilities should be assigned to node 1 and node n. Suppose $p_{11},\ldots,p_{1k}$ are assigned to node 1, and $p_{n1},\ldots,p_{nk}$ to node n. Then optimality requires

$$\left|\prod_{j=1}^{f} q_{1j} - \prod_{j=1}^{f} q_{nj}\right|$$

be minimized. However, such a balanced split is a well-known NP-complete problem except for $f = 2$. ∎

Another way to use the extra components is to construct f parallel consecutive-k systems. By a general result that component redundancy is better than system redundancy for any coherent system, the system obtained by pooling the f copies of node i from the f systems is more reliable than the parallel system. In fact the f-fold system and the f-parallel system provide the upper and the lower bound for the consecutive-fk-out-of-fn system.

THEOREM 9.3.3 $1 - \overline{R}(k,n)^f \le R(fk, fn) \le R(k,n,f)$.

Proof. Let $R(fk, fn)$ be obtained from the f-parallel system by linearly ordering the components into $v_{11}, v_{21}, \ldots, v_{f1}, v_{12}, v_{22}, \ldots, v_{fn}$. Then it is easily seen that every cutset of $R(fk, fn)$ is a cutset of f-parallel system. ■

THEOREM 9.3.4 *The optimal f-parallel system is induced by a consecutive partition, i.e., the best n components go to one system, the next n best go to a second system and so on.*

Proof. The f-parallel system can be viewed as a 2-stage system where the first stage is a parallel system with f subsystems. By a result of Hwang [106], any such system is optimized by a consecutive partition. ■

Let $S'(k, n, \{N_i\})$ denote a 2-stage system where the first stage is a consecutive-k system with n subsystems the ith of which is a series system with N_i components. Let $\sum_{i=1}^{n} N_i = N$.

THEOREM 9.3.5 *For given N_i, there exists an optimal $S'(k, n, \{N_i\})$ which is a consecutive partition.*

Proof. Hwang and Rothblum [114] showed that Theorem 9.3.5 is true for any coherent system whose last stage is series. ■

Theorem 9.3.5 reduces the search of an optimal system from comparing $N!/\prod_{i=1}^{n} N_i!$ systems to comparing $n!$ systems.

COROLLARY 9.3.6 *For $S'(2, n, \{N_i\})$, a best system is to follow the ordering of the optimal consecutive-2 system in assigning the consecutive sets of components.*

Proof. Suppose $i \prec j$ in the optimal consecutive-2 order. Then $I_i^b \prec I_j^b$. Since the product of the reliabilities of node i and j is a constant, we should maximize the reliability of node j. ■

We further generalize the model by assuming that the set of components is not fixed. Rather, the product of their unreliabilities is fixed at Q. The justification of this assumption is that suppose $q = f(c)$, where c is the cost of the component. Then $\ln f$ must be linear in c since we can reduce q to q^r by spending rc cost to install a component with r-redundancy. Hence $q = \alpha e^{-\beta c}$. But $f(0) = 1$. So $\alpha = 1$ and $q = e^{-\beta c}$. Consequently, with a fixed budget C, $\prod_{i=1}^{n} q_i = e^{-\beta C}$ is a constant.

Chang and Hwang [40] considered the continuous model where q_i can take any real value between 0 and 1. A *basic assignment* is one where all components are assigned to nodes in a minimum-k-cover set.

THEOREM 9.3.7 *There exists an optimal Q-line which is basic. For $n \geq k$ and $0 < p_i < 1$, every optimal line is basic.*

Proof. For $n \geq k$, $R \leq 1-Q$ since the system must fail if all components fail. For $k \leq n < 2k$, a basic assignment has reliability $1 - Q$, hence is optimal. We prove the general n case by induction. For $n \geq 2k$, it is better to reassign any component assigned to the last $k-1$ node to node $n - k + 1$ since any cutset of the latter is a cutset of the former. Now the n-line works if and only if node $n - k + 1$ works and the remaining $(n - k)$-line works. By induction, the $(n - k)$-line is basic. Hence the n-line is basic. ■

Let $m = \lfloor n/k \rfloor$ denote the size of a minimum-cover.

THEOREM 9.3.8 *For $n \geq k$, the basic system with the nonzero $q_i = Q^{1/m}$ is optimal with reliability $(1 - Q^{1/m})^m$.*

Proof. $R(L_n) = \prod_{i \in S} (1 - q_i) = \prod_{i \in S} (1 - e^{t_i})$, where $t_i = \ln q_i$.

Since R is symmetric in t_i and

$$(t_i - t_j)\Big(\frac{\partial R}{\partial t_i} - \frac{\partial R}{\partial t_j}\Big) = (t_i - t_j) R(L_n) \Big(\frac{-e^{t_i}}{1 - e^{t_i}} - \frac{-e^{t_j}}{1 - e^{t_j}}\Big) \leq 0,$$

is Schur convex and achieves maximum at $t_i = t = Q^{1/m}$. ■

Suppose $\gcd(n, k) = d$. Define a basic cycle as one which has nonzero reliabilities on every d-th node. For a given cycle C, let C_j denote the cycle obtained from C by reassigning all components to the next node congruent to j (mod d).

LEMMA 9.3.9 *$R(C) \leq R(C_j)$ for all j and the inequality is strict for the general case.*

Proof. It is easily verified that if C_j fails, then C also fails. ■

THEOREM 9.3.10 *For $d = k$, the basic cycle with $q_i =$ either 1 or $Q^{k/n}$ is Q-optimal, and uniquely Q-optimal in the general case.*

Proof. For $k|n$, n/k is the size of a minimum cover for both line and cycle. Since $(1 - Q^{k/n})^{n/k}$ is an upper bound of line, it is also an upper bound of cycle. But the basic cycle achieves this upper bound. It is also easily verified that for general p_i, no nonbasic cycle can achieve this bound. ■

By Lemma 9.3.9, we only need be concerned with basic cycles. In fact we can ignore all the 1-nodes and reduce the problem to $C(k' = k/d, n' = n/d)$ with $\gcd(k', n') = 1$.

Conjecture. Suppose $\gcd(k, n) = 1$. Then $q_i = Q/n$ is Q-optimal.

The conjecture is true for $k = 1, n - 1$.

THEOREM 9.3.11 *For $d = 1$, the uniform cycle is uniquely Q-optimal if $q_i \to 0$.*

Proof. When $q_i \to 0$, the goal is to minimize $f(C) = \sum_{i=1}^{n} \prod_{j=i}^{j+k-1} q_j$. Define

$$F(C) = f(C) - \lambda\left(\prod_{x=1}^{n} q^x - Q\right).$$

Then

$$\frac{\partial F}{\partial q_x} = \frac{1}{q_x}\left(\sum_{i=x-k+1}^{x} \prod_{j=i}^{i+k-1} q_j - \lambda Q\right)$$

or $\sum_{i=x-k+1}^{x} \prod_{j=i}^{i+k-1} q_j = \lambda Q$ is a constant. This implies

$$\sum_{i=x-k+1}^{x} \prod_{j=i}^{i+k-1} q_j = \sum_{i=x-k}^{x-1} \prod_{j=i}^{i+k-1} q_j$$

or

$$\prod_{j=x}^{x+k-1} q_j = \prod_{j=x-k}^{x-1} q_j \text{ for all } x.$$

Since $d = 1$, the above equality says that the product of any d consecutive reliabilities is a constant. Necessarily, $q_i = Q/n$. ■

In the discrete model we have $N > n$ components of common reliability q to be assigned to the n nodes. A variation is that these N copies are redundant units to be added to nodes already possessing reliabilities $\{p_i\}$. If the extra units have identical reliabilities, we can assign them one by one, each time to the node with currently highest Birnbaum importance. But if the units have distinct reliabilities, then the sequencing of the units to be assigned will affect the end result. To determine an optimal sequencing is an open problem. We could start by doing some simulation on two monotone sequence, one from large to small, and the other from small to large. If we have to start from an empty sequence, we need an initial assignment such that the system reliability is positive. One way is to find a minimum k-cover set and assign the m largest reliabilities one each to the m nodes in the cover set. Larger reliabilities should go to the more important units (structure-wise). For noncomparable units, we could use a random assignment or other methods.

9.4 WEIGHTED CONSECUTIVE SYSTEMS

Wu and Chen [257] generalized the consecutive systems to the weighted consecutive system by associating with component i a positive integer weight w_i such that the system fails if and only if there exists a set of consecutive nodes with total weight at least k all failed. The ordinary consecutive system is the special case $w_i = 1$ for all i.

Let $\widetilde{C}$ be the set of minimal cutsets. Wu and Chen gave an $O(n)$-time algorithm to compute $\widetilde{C}$. The algorithm starts from node 1 and adds the weights one by one of nodes $1, 2, \ldots$ until a sum $K \geq k$ is reached, say, at node j. Then subtracting from K the weight one by one of nodes $1, 2, \ldots$ until a further subtraction would reduce K to below k. Suppose i is the new beginning node. Then $(i, i+1, \ldots, j)$ is the first minimal cutset. Repeat the procedure by starting at node $i+1$ to pick up the other minimal cutsets.

Clearly, each minimal cutset can be computed in $O(n)$ time. Since there can be $O(n)$ minimal cutsets, it could take $O(n^2)$ time to compute $\widetilde{C}$. However, with more careful data processing, $\widetilde{C}$ can be computed in $O(n)$ time. The trick is not to do the addition each time from scratch. Suppose $K \geq k$ is the weight of the minimal cutset beginning with i and ending with j. Compute $K' = K - w_i$ and add to K' $w_{j+1}, w_{j+2}, \ldots$. This way, each weight w_i is processed at most two times in computing $\widetilde{C}$: once in addition, and once in subtraction. Since there are only n weights, $\widetilde{C}$ can be computed in $O(n)$ time.

Label $\widetilde{C}_i \in \widetilde{C}$ in the order they were identified in the above procedure. Let a_i and b_i denote the starting node and the ending node of the ith minimal cutset. Clearly $a_1 < a_2 < \ldots < a_m$ and $b_1 < b_2 < \ldots < b_m$, where $m = |\widetilde{C}|$.

THEOREM 9.4.1 *$x \in \widetilde{C}_i \cap \widetilde{C}_j$ implies $x \in C_h$ for all $i < h < j$.*

Proof. Suppose to the contrary that $x \in \widetilde{C}_i \cap \widetilde{C}_j$, but $x \notin C_h$, where $i < h < j$. Then either $x < a_h$ or $x > b_h$. In the first case $x < a_h < a_j$; then x is not in $\widetilde{C}_j$. IN the second case $x > b_h > b_i$; then x is not in $\widetilde{C}_i$. In either case we derive a contradiction to $x \in \widetilde{C}_i \cap \widetilde{C}_j$. ■

We can use the MCCS method of Hwang and Yao [119] to obtain an $O(kn)$-time algorithm of computing the reliability of the weighted system. Wu and Chen showed that it can be done faster. Clearly, $R(1,n) = R(1,b_m)$. They gave a recursive equation on $R(1,b_m)$.

THEOREM 9.4.2 $\overline{R}(1,b_m) = \overline{R}(1,b_{m-1}) +$

$$\sum_{i=0}^{a_m - a_{m-1} - 1} \left(\left(1 - \overline{R}(1, a_{m-1} + i - 1)\right) p_{a_{m-1}+i} \prod_{j=a_{m-1}+i+1}^{b_m} q_j \right).$$

Proof.

$$\begin{aligned}
\overline{R}(1,b_m) &= \Pr(\widetilde{C}_1 \cup \widetilde{C}_2 \cup \ldots \cup \widetilde{C}_m) \\
&= \Pr(\widetilde{C}_1 \cup \widetilde{C}_2 \cup \ldots \cup \widetilde{C}_{m-1}) + \Pr(\overline{\widetilde{C}_1 \cup \ldots \cup \widetilde{C}_{m-1}} \cap \widetilde{C}_m) \\
&= \overline{R}(1,b_{m-1}) + \prod_{\ell=a_{m-1}-1}^{a_m-2} \left(R(1,\ell) p_{\ell+1} \prod_{j=\ell+2}^{b_m} q_j \right) \\
&= \overline{R}(1,b_{m-1}) + \sum_{i=0}^{a_m - a_{m-1} - 1} \left(\left(1 - \overline{R}(1, a_{m-1} + i - 1)\right) \right. \\
&\qquad \left. p_{a_{m-1}+i} \prod_{j=a_{m-1}+i+1}^{b_m} q_j \right).
\end{aligned}$$

■

In computing $\overline{R}(1,b_1), \overline{R}(1,b_2), \ldots, \overline{R}(1,b_m)$, the total number of terms to be added is

$$(a_2 - a_1) + (a_3 - a_2) + \ldots + (a_m - a_{m-1}) = a_m = O(n).$$

Therefore $\overline{R}(1,b_m)$ can be computed in $O(n)$ time.

Chang, Chen and Hwang [46] gave the circular version. Define $T = \max\{i : \sum_{j=1}^{i-1} w_j < k, 1 \le i \le n+1\}$. Set $w_k = 0, w_k = w_{k-n}$ for $k > n$. Let B_i denote the event that the working component with minimum index is i.

THEOREM 9.4.3

$$R_C(1,n) = \begin{cases} 1, & \text{if } T > n, \\ \sum_{i=1}^{T} \left(R_L(i+1, n+i-1) \left(\prod_{j=1}^{i-1} q_j \right) p_i \right), & \text{if } T \le n. \end{cases}$$

Proof. The $T > n$ case is obvious. If $T \le n$, a working system must contain at least one working component. In other words, at least one of $B_1, B_2, \ldots, B_n$ must occur, and these events are mutually exclusive. Let S denote the event that the system is working.

$$\begin{aligned} R_C(1,n) &= \sum_{i=1}^{n} \Pr(S|B_0)\Pr(B_i) \\ &= \sum_{i=1}^{T} \Pr(S|B_i)\Pr(B_i) \\ &= \sum_{i=2}^{T+1} \left(R_L(i, n+i-2) \left(\prod_{j=1}^{i-2} q_j \right) p_{i-1} \right) \\ &= \sum_{i=1}^{T} \left(R_L(i+1, n+i-1) \left(\prod_{j=1}^{i-1} q_j \right) p_i \right). \end{aligned}$$

■

It takes $O(Tn)$ time to compute $R_C(1,n)$, while T is bounded by $\lceil (k - w_{\max})/w_{\min} \rceil - 1$.

Chapter 10

APPLICATIONS

We collect applications of the consecutive system from the literature to illustrate the models and the results discussed in previous chapters. The first section would be restricted to the modeling, while the second section contains reliability computations.

10.1 EXAMPLES OF MODELINGS

Example 1. The Telecommunication and Oil Pipeline System

Consider a telecommunication system with n relay stations. We will name stations consecutively from 1 to n. Suppose a signal emitted from station 1 can be received by both stations 2 and 3, and a signal relayed from station 2 can be received by both stations 3 and 4, etc. Thus, when station 2 is failed, the telecommunication system is still able to transmit a signal from station 1 to station n. However, if both stations 2 and 3 are failed, then a signal cannot transmit from station 1 to station 4, and the system fails. Similarly, if any two consecutive stations in the system fail, the system fails.

Another example is an oil pipeline system with n pump stations, where each pump can send flow to the next two stations. If one pump station is down, the flow of oil would not be interrupted because the neighboring stations could carry the load. However, when two consecutive pump stations are down, then the oil flow is stopped and the system fails.

Both examples illustrate the consecutive-2-out-of-n system, because any two consecutive components constitute a cutset. They could be extended to illustrate the consecutive-k-out-of-n system.

Example 2. The Street Parking Problem

Street parking is an example of the consecutive-2-out-of-n:G system. Suppose that there are n parallel parking spaces on a street and each space is suitable for one car. If a bus parks on the street, it takes two consecutive spaces. Every parking space has a probability of being available. The problem of interest is to find the probability that the bus can park on this street, i.e., at least two consecutive parking spaces on the street are empty.

In practice, the assumption of independent states is untrue in parallel parking. But for simplicity the system reliability can be approximated under the IID or the IND model.

Example 3. Quality Control

In lot acceptance sampling, a lot can be rejected if it contains k consecutive defectives. The criterion can be relaxed to require b occurrences of k consecutive defectives, or an occurrence of k defectives out of m consecutive samples. In practice, a more sophisticate sampling plan may be used. If consecutive k lots are rejected under the normal sampling scheme, a tightened sampling scheme becomes effective. If consecutive k' lots are then accepted (a consecutive-k'-out-of-n:G system is working) under the tightened sampling scheme, the normal sampling scheme returns. Otherwise, the inspection is discontinued and the products are rejected.

Another example to use the k-within-consecutive-m-out-of-n system in quality control is the Shewhart control chart. The basic idea is to initiate a search for a cause of shift in a production line, whenever k points among m consecutive points fall outside the warning limits.

Example 4. Pattern Detection

The presence of a disease is diagnosed by an X-ray; p is the probability that an individual cell (or other small portion of the X-ray) is healthy. Unless diseased cells are aggregated into a sufficiently large pattern (say, a $k \times k$ square) the radiologist might not detect their presence. Let $P_f(k,n)$ be the detection probability of a k-square. In general we want to determine the probability of detecting patterns of arbitrary shapes and sizes. For an arbitrary pattern T, let k_1 and k_2 be the sizes of the largest inscribed square in T and smallest circumscribed square about T respectively, and let $P_f(T)$ be the probability of detecting pattern T. Then

$$P_f(k_2,n) \le P_f(T) \le P_f(k_1,n).$$

This is a two-dimensional consecutive-k system.

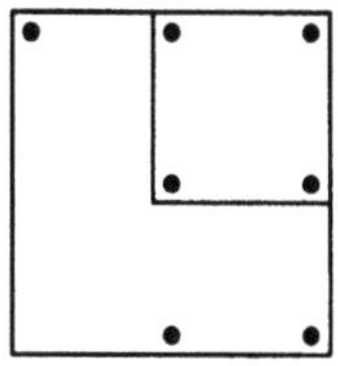

Figure 10.1.1. Detection pattern.

For instance, the pattern of Figure 10.1.1 has a detection probability between $P_f(3,n)$ and $P_f(2,n)$.

Example 5. Power Supply Lines

Consider Figure 10.1.2 in the following, where the electricity is supplied through a total of six electric power lines.

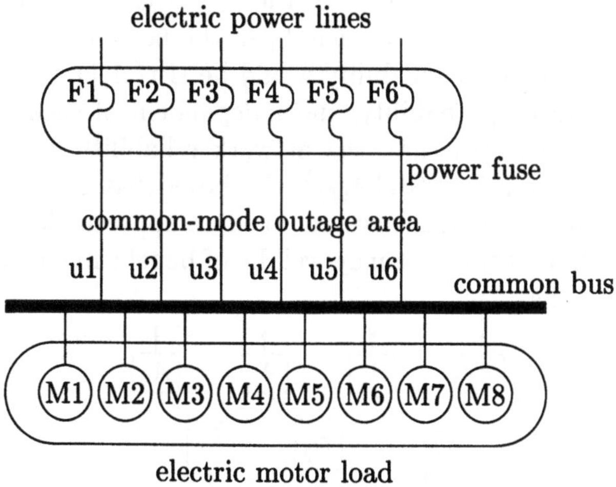

Figure 10.1.2. Power supply line digram.

Consider a pattern where components fail due to common-mode forced outages in which a single external cause results in multiple component failures. In order to prevent the overcurrent caused by, say, a thunder, a fuse is installed on each power supply line that is connected to the common bus. Here we assume that at least four power lines should not be failed consecutively in order to continue the power supply, that is, $k = 4$. When less than four fuses are consecutively disconnected by the common-mode outage events, the power can be supplied. Otherwise, the power supply becomes zero.

Example 6. Deteriorating Amplifiers

An example of a linear consecutively-connected system with multi-state components is as follow. The components of the linear system are a set of relay stations; the source is a transmitter station; and the sink is a receiver. Each of the $n+1$ stations C_i consists of k_i amplifiers. If all the amplifiers of a C_i are operating, then a signal from C_i reaches the next $\min\{k_i, n-i+1\}$ stations. The failure of one amplifier reduces the range of C_i by 1 station, i.e., a signal from that station reaches the next $k_i - 1$ stations, and so on.

Example 7. The Network Design

For the network shown in Figure 10.1.3, suppose component B_i (for all i) is perfect. Each B_i, upon the arrival of a signal split and send the signal to A_i and A_{i+1}. Then A_i sends out the corresponding signals. In this way, redundancy of the B_i signal is introduced through a consecutive-2 structure. If each B_i in the network was followed by a single A_i, the network reliability would be $(\Pr\{A_i\})^3$ (assume every A_i has same working probability, and independent of each other). Under the consecutive-2 structure, the network reliability is Pr{consecutive-2-out-of-4:F system} $\geq (\Pr\{A_i\})^3$. It shows that the redundancy and increased complexity of the system yield an appreciable gain in reliability. Calculation of this sort may well be of benefit in cost and reliability design trade-offs.

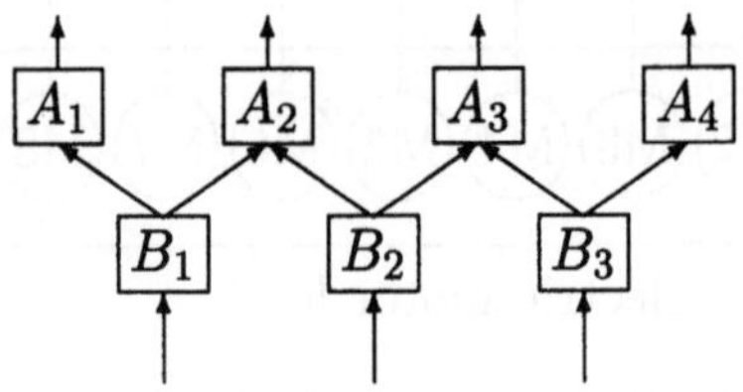

Figure 10.1.3. A network with the consecutive-2 structure.

Example 8. The Supervision System

Consider a supervision system as shown in Figure 10.1.4. The nodes indicate the camcorders. Every camcorder supervises the area with radius r, and all camcorders are same. The system fails if and only if some area in a $3r \times 3r$ rectangle are out of supervision. So the system fails if and only if there are at least two adjacent camcorders failed, e.g. the black area in Figure 10.1.4 is out of supervision. This is a 2-dimensional consecutive system.

Figure 10.1.4. Working/failed system.

Example 9. The Measure System

A cylinder body (for example, reaction chamber) surface is covered by temperature sensors, the sensor system fails if and only if all sensors in a consecutive area fail. This is a 2-dimensional consecutive system. The system is equivalent to the following 2-dimensional consecutive system.

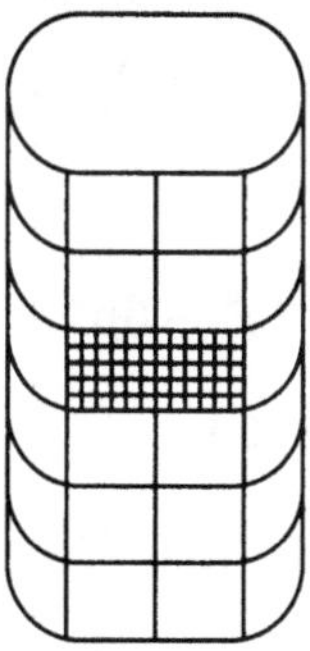

Figure 10.1.5. Temperature sensor digram.

10.2 APPLICATION EXAMPLES WITH COMPUTATIONS

Example 1. Satellite Communication System

Suppose that there are three ground communication stations and three connectors in a satellite, which consist of a simple circular communication system. Every ground communication station is covered by two connectors in the satellite and every connector is connected to two ground communication stations. See Figure 10.2.1 for the system.

The reliabilities of the communication links (including satellite and ground receiving equipment) are p_i ($1 \leq i \leq 6$). The communication

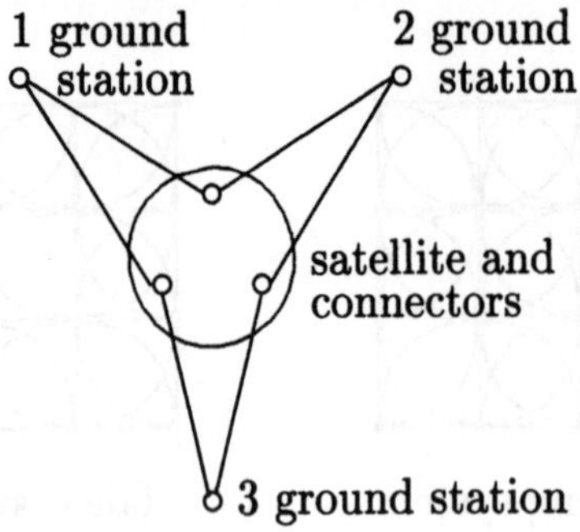

Figure 10.2.1. Satellite communication system.

system reliability is the probability that no ground communication station or connector is cut off. Note that the system fails if and only if two consecutive links fail in the cycle of six links.

Using the inclusion and exclusion formula, we have

$$\begin{aligned} R &= p_1p_3p_5 + p_2p_4p_6 + p_1p_3p_4p_6 + p_1p_2p_4p_5 + p_2p_3p_5p_6 - \\ & p_2p_3p_4p_5p_6 - p_1p_2p_3p_5p_6 - p_1p_3p_4p_5p_6 - p_1p_2p_3p_4p_6 - \\ & p_1p_2p_3p_4p_5 - p_1p_2p_4p_5p_6 + 2p_1p_2p_3p_4p_5p_6. \end{aligned}$$

If $p_i \equiv p$ $(1 \le i \le 6)$, we have

$$R = 2p^3 + 3p^4 - 6p^5 + 2p^6,$$

which is increasing in p for $0 < p < 1$.

Example 2. Tolerance for Satellite Computer MOS Memory

The satellite computers are very important for satellite working in the space. For consideration of quality, long working time, need of correction for operation error and etc., redundancy for parts are usually built in for satellite computers. A satellite computer requires its 64K memory's consecutive 32K digits working normally for ensuring 16-bit computer working. This is a linear consecutive-k-out-of-n:G system ($n = 2k$). Suppose that the life time of units of memories are exponentially distributed (independently), and 16 digits as one unit has a failure rate $\lambda = 0.64$fit (1fit $= 10^{-9}$/hour). The question is, when the satellite mission time is 8 years, what is the reliability of the satellite computer memory?

As $1K = 1024$ digits, 1K consists of $1024/16 = 64$ units. The satellite computer memory consists of $64 \times 64 = 4096$ units, so our problem is a large linear consecutive-2048-out-of-4096:G system ($n = 4096$ and $k = 2048$). The memory unit reliability in 8 years as

$$p = e^{-\lambda t}, \text{ where } t = 24 \times 365 \times 8 \text{ hours, i.e.,}$$

$$p = e^{-0.64\times 10^{-9}\times 24\times 365\times 8} = 0.999955.$$

On the basis of linear consecutive-k-out-of-n:G system formula, we have the satellite computer memory reliability in 8 years is

$$\begin{aligned} R &= p^k[1+(n-k)(1-p)] \\ &= 0.99955^{2048}[1+(4096-2048)(1-0.999955)] \\ &= 0.996003. \end{aligned}$$

Example 3. Relayed Mobile Communication System

Chiang and Chiang [60] considered a relayed mobile communication system with evenly spaced mobile relay stations (spacecraft) moving at the same speed from an origin (earth) toward a destination (far away star). The background for their problem was an attempt to improve our understanding of Saturn by the data sent back from Voyager 1 in 1980 November. Recall that the first encounter with Saturn by Pioneer 11 fourteen months earlier in 1979 September failed in sending back data. The Pioneer 11 was not lost in space. There are two possible explanations for the loss of contact with Pioneer 11: 1) The spacecraft was no longer functioning, or 2) the spacecraft was too far away from the earth so that the ground tracking stations were not able to pick up signals sent by the Pioneer. In order to explore distant stars, one approach is to launch a second spacecraft before ground stations lose contact with the first spacecraft. The second spacecraft will relay signals transmitted by the first spacecraft to ground stations; of course, additional spacecraft will be launched if needed to relay signals from the first spacecraft. If spacecraft i fails, signals from the first $i-1$ spacecrafts are lost, and spacecraft $i+1$ is now the first spacecraft to explore the distant star. An alternative is to launch the space crafts in shorter intervals so that signals transmitted by a spacecraft can reach their next k spacecrafts. Suppose that the constant distance between two successive spacecrafts is represented as $1/D$ of the distance from the star to the earth where D is an integer. The question is, "What is the mean number of spacecrafts needed to be launched to get a successful transmission?" The above problem can be modeled by an infinite window system which works if a good window exists where a window is a consecutive-k-out-of-D line.

Chiang and Chiang derived the mean number of spacecrafts needed to accomplish the mission as follows: Let

N be the number of relay stations needed for the mission to succeed;

p be the reliability of a relay station;

$R(k, kD; p)$ be the reliability of a consecutive-k-out-of-kD:F system with reliability p for each component;

Δ be the event "first system failure occurs at stations $x+1,\ldots,x+k$". Then

$$E_k\{N\} = E_x\{E_k\{N \mid \Delta\}\},$$

$$E_k\{N|\Delta\} = \begin{cases} (x+k)+E_k\{N\}, & \text{if } 0 \le x \le kD-k; \\ kD, & \text{otherwise,} \end{cases}$$

$$\Pr\{\Delta\} = \begin{cases} q^k, & \text{if } x=0; \\ R(k,x-1;p)pq^k, & \text{if } 0 \le x \le kD-k; \\ R(k,kD;p), & \text{otherwise,} \end{cases}$$

where $R(k,n;p) = 1$ if $k > n$. Substitute the second and the third equations into the first equation, we get

$$\begin{aligned} E_k\{N\} &= \sum_x E_x\{N|\Delta\}\Pr\{\Delta\} \\ &= [k+E_k\{N\}]q^k + kDR(k,kD;p) + \\ &\quad \sum_{x=1}^{kD-k} (x+k+E_k\{N\})R(k,x-1;p)pq^k. \end{aligned}$$

Factor out $E_k\{N\}$ and note that

$$q^k + \sum_{x=1}^{kD-k} R(k,x-1;p)pq^k + R(k,kD;p) = 1,$$

we get

$$E_k\{N\} = \frac{kq^k + kDR(k,kD;p) + \sum_{x=1}^{kD-k}(x+k)R(k,x-1;p)pq^k}{R(k,kD;p)}.$$

Example 4. The Railroad Operation in a Station

For the linear consecutive-k-out-of-n:G system, Zhang, Miller and Kuo [264] presented a very good application example in railroad operation. Their results can help in scheduling the incoming trains to utilize the station resources better. The case is as follows:

A railroad system has 17 lines numbered from line 1 to line 17. The first nine lines constitute the basic section that receives and sends trains, and the remaining eight lines serve as the assembly section that organizes or reorganizes trains. The utilization density of a line can be considered as the probability that the line is not available. All lines in the basic section have the same utilization density of $u = 0.35$, and all lines in the assembly section have the utilization density of $u = 0.5$.

The dispatcher has been informed that a special train is coming. However, because of over-limit loading of some vehicles, the neighbor lines of the line that receives the train must be empty; that is, it is required that at least three consecutive lines are empty so that the train can arrive. In addition, because of physical limitations, lines 1, 9, 10 and 17 cannot be used to receive the train. What is the probability that the special train can enter the station without delay, given that the assembly section can also accept the special train?

The problem, in fact, can be formulated as the reliability problem of a linear consecutive-k-out-of-n:G system. Taking into account the restrictions imposed on the stations, we can regard the stations as a linear consecutive-3-out-of-8:G system with a dummy line between lines 9 and 10. Obviously, the dummy line assumes the utilization density of $u = 1$.

In summary,

$$u_1 = u_2 = \ldots = u_9 = 0.35, u_{10} = 1, u_{11} = \ldots = u_{18} = 0.5.$$

The utilization density of a line in this problem is similar to the unreliability of a component of a reliability system. Since $u_{10} = 1$, the problem is to find the conditional probability that the system will work, that is, to find

$$\Pr\{\text{accept the train} \mid \text{lines 9 and 10 cannot be used}\} = R(18; 3 \mid u_{10} = 1).$$

Actually, we can find the probability, respectively, for a linear consecutive-3-out-of-9:G system (for lines 1 through 9) and for a linear consecutive-3-out-of-8:G system (for lines 10 through 17). Then the probability of interest is obtained as follows:

$$\begin{aligned} R(18; 3 \mid u_{10} = 1) &= R(9; 3) + R(8; 3) - R(9; 3)R(8; 3) \\ &= 0.744431 + 0.417969 - 0.744431 \times 0.417969 \\ &= 0.8512509. \end{aligned}$$

The Birnbaum importance of components for lines in basic section and assembly section is given in the following table.

Table 10.1: Birnbaum importance of lines in basic section and assembly section.

basic section		assembly section	
line	importance	line	importance
1	0.07883745	10	0.10156250
2	0.17188859	11	0.21093750
3	0.27915323	12	0.33593750
4	0.20177740	13	0.25781250
5	0.21599090	14	0.25781250
6	0.20177740	15	0.33593750
7	0.27915323	16	0.21093750
8	0.17188859	17	0.10156250
9	0.07883745		

From the results in the above table, we know that lines 3, 7, 12 and 15 are the most important ones in the corresponding systems. If these lines are not available at the time the train comes, the probability of permitting the train to enter the station will be greatly reduced. For instance, if line 3 is not empty, then lines 1 through 4 cannot be used to receive the train. However, if line 1 is not empty, only line 2 cannot be used. Reliability importance increases from line 1 to line 3 and decreases from line 7 to line 9 (last three lines) in the basic section.

The optimal assignment was studied under the assumption that utilization densities are not the same.

From above we know that on the average, about $0.35 \times 9 = 3.15$ trains stay in the basic section at any time. If we can control the utilization density for each line by changing schedules of train operation, optimal assignment of utilization density for all lines may take place.

Suppose that we have nine utilization densities to be assigned to nine lines in the basic section:

$$u_i = 0.1 + 0.05i \quad \text{for } i = 1, 2, \ldots, 9, \quad \sum_{i=1}^{9} u_i = 3.15,$$

which represents the fact that about 3.15 trains are standing in the basic section at any time. Equivalently, we have

$$p_{10-i} = 0.9 - 0.05i \quad \text{for } i = 1, 2, \ldots, 9,$$

that is,

$$p_1 < p_2 < \cdots < p_9,$$

where $p_{10-i} = 1 - u_i$ represents the probability that the line of interest is empty. In this problem, we are interested in the assignment maximizing the probability that the special train enters the basic section of the station without delay. That is, we want the optimal configuration of a linear consecutive-3-out-of-9:G system.

Intuitively, we should assign less reliable components to the end positions of a linear consecutive-k-out-of-n:G system, and more reliable components should go to the middle positions of the system, since the middle positions produce more consecutive-k-tuples of components than the end positions.

For our problem of the linear consecutive-3-out-of-9:G system, $9!/2 = 181440$ possible configurations exist. Based on their best information, 12 configurations are listed in the following table with corresponding system reliabilities. Each configuration is selected in such a way that the four largest utilization densities are assigned to lines 1, 2, 8 and 9, and lines 3 and 7 are assigned by either the largest or the smallest ones among the remaining densities. The last three densities are assigned enumeratively to compare what assignment generates the best configuration. Configuration 8 (starred) gives the best system reliability among 12 configurations from following table in this case.

Table 10.2: Configurations of the linear consecutive-3-out-of-9:G system.

configuration number	configuration	system reliability
1	$(1,3,8,5,6,7,9,4,2)$	0.81724846
2	$(1,3,8,5,7,6,9,4,2)$	0.81774491
3	$(1,3,8,6,7,5,9,4,2)$	0.81636608
4	$(1,3,8,6,5,7,9,4,2)$	0.81372321
5	$(1,3,8,7,5,6,9,4,2)$	0.81236666
6	$(1,3,8,7,6,5,9,4,2)$	0.81451291
7	$(1,3,5,7,8,9,6,4,2)$	0.82784170
8*	$(1,3,5,7,9,8,6,4,2)$	0.82924467
9	$(1,3,5,8,9,7,6,4,2)$	0.82803583
10	$(1,3,5,8,7,9,6,4,2)$	0.82386810
11	$(1,3,5,9,7,8,6,4,2)$	0.82267875
12	$(1,3,5,9,8,7,6,4,2)$	0.82544333

Therefore the assignment should be

$$\begin{aligned} p(1) &= p_1, & p(4) &= p_7, & p(7) &= p_6,\\ p(2) &= p_3, & p(5) &= p_9, & p(8) &= p_4,\\ p(3) &= p_5, & p(6) &= p_8, & p(9) &= p_2, \end{aligned}$$

where $p(i)$ is the probability that line i is empty. This assignment maximizes the probability that the special train enters the basic section without delay.

If lines 1 and 17 can be used to receive the special train, what is the probability that the special train can enter the station?

It is clear that the problem can also be formulated as the reliability of a linear consecutive-3-out-of-20:G system with three dummy lines where one is between lines 9 and 10, with $u_{10} = 0$, the other two are outside of lines 1 and outside of line 17, with $u_0 = u_{10} = 1$. Similarly,

$$\begin{aligned}&\Pr\{\text{accept the train} \mid \text{lines 9 and 10 can not be used}\}\\ &= R(19; 3 \mid u_0 = 0, u_i = 0.35\ (1 \le i \le 9), u_{10} = 1,\\ &\qquad\qquad u_j = 0.5\ (11 \le j \le 18), u_{19} = 0\},\end{aligned}$$

and the probability is obtained, respectively, for a linear consecutive-3-out-of-10:G system (for lines 1 through 9 and dummy line 0) and a linear consecutive-3-out-of-9:G system (for lines 10 through 17 and dummy line 18). The probability of interest is

$$\begin{aligned}&R(20; 3 \mid u_0 = u_{19} = 0, u_{10} = 1)\\ &= R(10; 3 \mid u_0 = 0) + R(9; 3 \mid u_{19} = 0)\\ &= R(9; 3) + 0.35 \times 0.65^2 + R(8; 3) + 0.35 \times 0.5^2 -\\ &\quad [(R(9; 3) + 0.35 \times 0.65^2)(R(8; 3) + 0.35 \times 0.5^2)]\\ &= 0.744431 + 0.35 \times 0.65^2 + 0.417969 + 0.35 \times 0.5^2 -\\ &\quad [(0.744431 + 0.35 \times 0.65^2)(0.417969 + 0.35 \times 0.5^2)].\\ &= 0.946742.\end{aligned}$$

Example 5. Optimal Design for Camera Monitoring System

Shen and Zuo [241] presented an application of consecutive-k-out-of-n:G system. The case is as follows:

The monitoring problem of a nuclear reaction in a reactor will be considered. To obtain 3-dimensional snap shots of the reactions in a certain area, say A, in the reactor, four high speed cameras are mounted at slightly different angles focusing on the area of interest (see Figure 10.2.2).

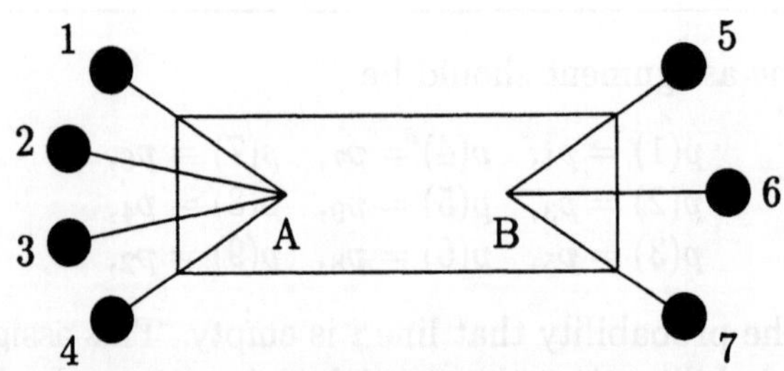

Figure 10.2.2. A monitoring system with seven cameras.

The four cameras are labeled 1, 2, 3 and 4. At least two adjacent cameras have to work properly for us to obtain the required data or image regarding the reactions in the area. Thus the four cameras form a linear consecutive-2-out-of-4:G system. For the second area of interest, say, area B, three cameras (labeled 5, 6 and 7) are available to monitor this area. If we also need at least two adjacent cameras to work in order to obtain an acceptable image of the reactions in area B, we have a linear consecutive-2-out-of-3:G system. If we need high quality snapshots of both areas in order to analyze the reactions in the reactor, then the linear consecutivr-2-out-of-4:G system with cameras 1,2,3, and 4 and the linear consecutive-2-out-of-3:G system with cameras 5,6,and 7 are functionally connected in series. If we have seven cameras with different reliabilities available, how should we arrange these cameras to achieve the highest system reliability?

Note that, the camera system described above may also be used in analysis of fluid dynamics in pipeline systems and flame propagation in engines and fires.

For the camera system , we have a linear consecutive-2-out-of-4:G system functionally connected with a linear consecutive-2-out-of-3:G system in series. The linear consecutive-2-out-of-3:G system can be further decomposed into a series subsystem consisting of camera 6 and a linear consecutive-1-out-of-2:G subsystem (i.e. a parallel system) with cameras 5 and 7. Thus the system of cameras can be represented by the network diagram in the following.

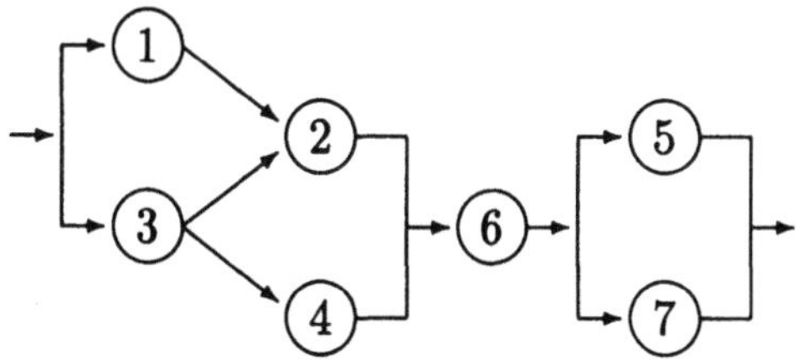

Figure 10.2.3. Diagram of camera system.

Suppose we have 7 cameras of different reliabilities available, i.e.,

$$p_{[1]} < p_{[2]} < p_{[3]} < p_{[4]} < p_{[5]} < p_{[6]} < p_{[7]}.$$

The optimal design of the system should assign $p_{[7]}$ to camera 6. It follows from Theorem 8.3.8 that the six cameras left, $p_{[6]}, p_{[5]}, p_{[2]}$ and $p_{[1]}$ should be assigned to the linear consecutive-2-out-of-4:G system and $p_{[3]}, p_{[4]}$ should be assigned to the linear consecutive-1-out-of-2:G system. An optimal assignment of $p_{[6]}, p_{[5]}, p_{[2]}$ and $p_{[1]}$ in the linear consecutive-2-out-of-4:G system is $p_1 = p_{[1]}, p_2 = p_{[5]}, p_3 = p_{[6]}, p_4 = p_{[2]}$ according to Corollary 3.1.6, $p_{[3]}$ and $p_{[4]}$ can be arbitrarily assigned to p_5 and p_7. We will let $p_5 = p_{[4]}, p_7 = p_{[3]}$. Thus , an optimal design of the system of cameras is:

$$p_6 > p_3 > p_2 > p_5 > p_7 > p_4 > p_1,$$

where p_i is the reliability of the camera labeled i in Figure 10.2.3 ($1 \leq i \leq 7$).

Example 6. Application to Computer Network

Hwang [103] discussed the simplified reliability for consecutive-k-out-of-n system and applied his results to compute the reliability of a k-loop computer network. The case is as follows:

Consider n stations denoted by the residues modulo n. A k-loop network for n stations consists of k loops with links $i \longrightarrow i + s_j, i = 0, 1, \ldots, n-1, j = 1, \ldots, k$. The k-loop networks have been widely studied as topologies for computer networks.

Assume that each station can fail independently with probability q but links always work. A common measure for reliability is to call the system failed if there exist two working stations A and B such that every path from A to B must go through a station which has failed. We now use the result of $R_L(k, n; p)$ and $R_C(k, n; p)$ to compute the reliability of a k-loop network with $s_i = i$.

We give an example for $n = 6, k = 2, s_i = i$, the network diagram is as in Figure 10.2.4.

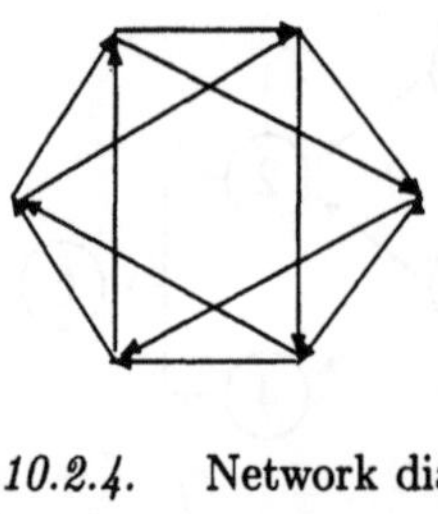

Figure 10.2.4. Network diagram.

Hwang [103] showed that the network fails if and only if it contains at least two working stations and k consecutive failing stations. Therefore the reliability of the network is simply the reliability of a consecutive-k cycle plus the reliability that at most one station works. We have:

The reliability of a k-loop network with $s_i = i$ and n stations is

$$q^{n-1} + \sum_{i \geq 0} (-pq^k)^i \binom{n-ki}{i} - q^n + k \sum_{i \geq 0} (-pq^k)^{i+1} \binom{n-k(i+1)-1}{i}$$

for $k \geq 1$.

Note that the above formula minus the q^{n-1} term is the reliability $R_C(k, n; p)$.

Example 7. Large Dependent Oil Pump Transporting System

Fu [77] discussed the following problem. Consider two systems for transporting oil from point A to point B by pressure pumps and pipelines. System #1 uses 30 α-type pressure pumps (equally spaced between A to B); each pump can move oil to the next pump. If any one of the 30 α-type pressure pumps breaks down then the system cannot transport the oil. The broken pressure pump needs to be repaired or changed immediately. System #2 is to replace the α-type pumps β-type pumps where each β-type pump is able to transport the oil a distance of 3 pumps. Therefore, the system fails if and only if 3 consecutive pressure pumps fail. Let the unreliability of the α-type pressure pump be 0.33% and the unreliability of the β-type pressure pump be 3.2% (10 times as bad). The upper bounds of unreliability for systems #1 and #2 are

$$\begin{aligned}
\Pr\{\text{system \#1 fails}\} &= F_L(n = 30, k = 1) \\
&\le 1 - (1 - q^k)^{n-k+1} \\
&= 1 - (1 - 0.0033)^{30} \\
&= 9.5\%, \\
\Pr\{\text{system \#2 fails}\} &= F_L(n = 30, k = 3) \\
&\le 1 - (1 - q^k)^{n-k+1} \\
&= 1 - (1 - 0.032^3)^{28} \\
&= 0.09\%.
\end{aligned}$$

From the above results, we know that the unreliability of system #1 is 100 times more than system #2. The cost of building system #1 might be less than the cost of building a consecutive-3-out-of-30:F system. However, the advantage of high reliability of consecutive-3-out-of-30:F system could easily far outweight the slightly higher cost. If repair costs and inconvenience are considered, the incentive for using the consecutive-3-out-of-30 system is very strong.

Example 8. Infrared Detection and Signal Processing System

Tung [246] presented a reliability analysis of a complex system such as infrared (IR) systems. In his paper, two examples were given: a) J2 detector/channel/preamplifer; b) nonlinear multidetector subsystem. In the space or military applications an IR system usually is a sophisticated and complicated equipment, especially the detecting and signal processing portions of the systems. Due to its complex wiring configuration and intricate failure criteria, a special treatment was taken to compute the system reliability. The system reliability diagram is as in Figure 10.2.5. Note that in the figure, D indicates detectors in sensor, the numbers indicate the number of channels, and MUX is MUX card.

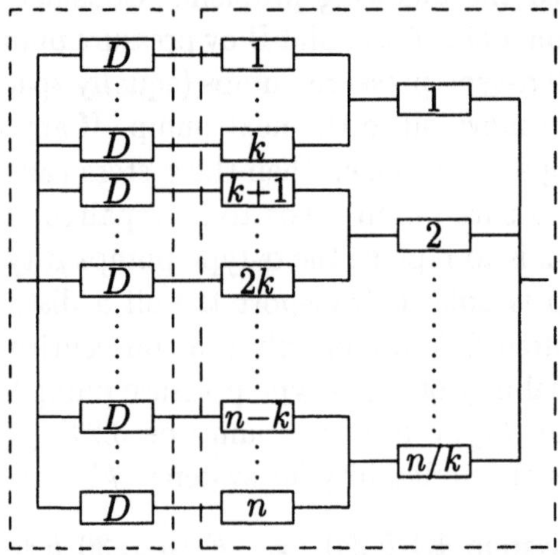

Figure 10.2.5. Detector/channel/preamplifer reliability diagram.

Sun and Liao [243] introduced the consecutive system concept to Tung's example [246].

For a detector/channel/preamplifer system consisting of 112 detectors/channels and 28 MUX cards, the failure criteria are the occurrence of any of the following conditions:

(1) more than five dead or noisy channels;

(2) three or more dead or noisy channels adjacent to one another;

(3) one or more dead or noisy channels in the central ten percent of the array.

Based on the reliability block diagram shown in the above and the failure criterion (2), no MUX card is permitted to fail since one MUX card failure will cause four adjacent channel failures. The criteria (1) and (3) allow five detector/channel failures outside the central ten percent of the array but no failure is allowed inside.

Another example of (n, f, k) system is a bank automatic payment system, which consists of n terminals and several MPX connected with center computer as shown in Figure 10.2.6.

The system failure is defined as the occurrence of any of the following conditions:

(1) more than f (exclusive) failed terminals,

(2) k or more consecutive failed terminals.

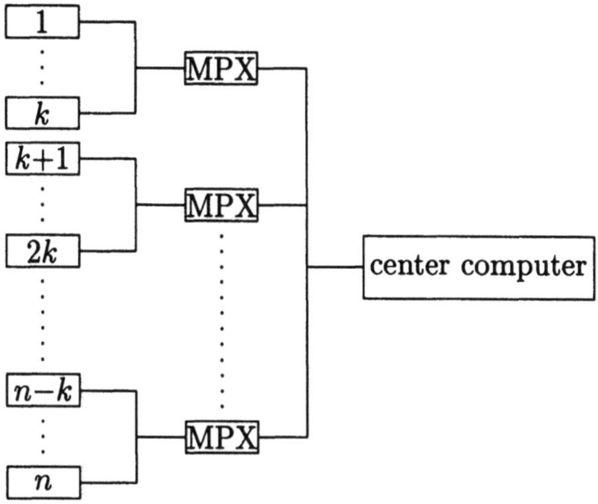

Figure 10.2.6. Bank automatic payment system.

If the above condition (1) or (2) happens, then at least one subbank cannot contact with the center computer or the whole system's payment speed decreases dramatically.

The system reliability is as follows.

Case 1, when $k = 1$, the (n, f, k) system becomes a series system, and so

$$R_{\text{sys}} = (1-\theta)^n,$$

where θ is detector/channel/terminal (for short, denoted by DCT) failure probability.

Case 2, when $k > f$, the (n, f, k) system becomes a consecutive-k system, and so

$$R_{\text{sys}} = \sum_{j=0}^{f} \binom{n}{j} \theta^j (1-\theta)^{n-j}.$$

Case 3, when $1 < k \leq f$,

$$\begin{aligned} R_{\text{sys}} &= \sum_{j=0}^{f} \Pr\{j \text{ DCTs fail in working system}\} \\ &= \sum_{j=0}^{k-1} \Pr\{j \text{ DCTs fail in working system}\} + \\ &\quad \sum_{j=k}^{f} \Pr\{j \text{ DCTs fail in working system}\} \\ &= \sum_{j=0}^{k-1} \binom{n}{j} \theta^j (1-\theta)^{n-j} + \Pr\{\text{consecutive-}k\text{-out-of-}n \end{aligned}$$

system works with j failed units for $k \leq j \leq f\}$

$$= \sum_{j=0}^{k-1} \binom{n}{j} \theta^j (1-\theta)^{n-j} + \sum_{j=k}^{f} N(j,n,k) \theta^j (1-\theta)^{n-j},$$

where $N(j, n, k)$ is the number of ways of arranging j failed components and $n - j$ working components into a working system without k consecutive components all failed.

As an example, for an $(n, f, k) = (8, 4, 3)$ system, the system reliability is

$$\begin{aligned} R_{\text{sys}} &= \sum_{j=0}^{2} \binom{8}{j} \theta^j (1-\theta)^{8-j} + \sum_{j=3}^{4} N(j,8,3) \theta^j (1-\theta)^{8-j} \\ &= (1-\theta)^8 + 8\theta(1-\theta)^7 + 28\theta^2(1-\theta)^6 + \\ &\quad N(3,8,3)\theta^3(1-\theta)^5 + N(4,8,3)\theta^4(1-\theta)^4 \\ &= (1-\theta)^8 + 8\theta(1-\theta)^7 + 28\theta^2(1-\theta)^6 + \\ &\quad 50\theta^3(1-\theta)^5 + 45\theta^4(1-\theta)^4. \end{aligned}$$

Note that, θ is usually a function of time t.

If we assume that repair is instantaneous and the maintenance policy is "good as new", that is, after maintenance, the whole system is like a new one. The MTBF (mean time between failure) can be determined for both maintenance and nonmaintenance cases as follows:

$$\text{MTBF} = \frac{\int_0^T R_{\text{sys}}(t)dt}{1 - R_{\text{sys}}(T)}.$$

When set $T = \infty$, the MTBF for nonmaintenance is obtained, where T is the interval maintenance time regardless whenever the system fails or not.

Example 9. Furnace System and Its Tubes Reliability

Zuo and Wu [272] presented an application example using k-out-of-n and consecutive-k-out-of-n system. The application is in a petrochemical company which produces crude oil and other chemical products. It uses several hydrogen plants to provide hydrogen for hydro-treating. The company's oil production rate is proportional to the amount of hydrogen supplied. Therefore, it is very important that the hydrogen plants operate without interruptions or equipment failures. Of specific interest to the company is the operation of the methane reformer furnaces. These furnaces have hundreds of tubes which are filled with catalyst. Methane and steam are passed through these tubes at high temperature and hydrogen is produced. The retubing, i.e., the replacement of all the tubes, costs in the order of ten million dollars. The tube arrangement in one of the furnaces is shown in Figure 10.2.7.

1	2	3	4	...	114
115	116	117	118	...	228
229	230	231	232	...	342
343	344	345	346	...	456

Figure 10.2.7. Tube arrangement diagram.

Since the working conditions (temperature and pressure) of the furnace tubes are quite different from the design conditions, it is natural to expect that the actual life of the tubes will be different from the design life. To prevent replacing the tubes too early or too late, the remaining useful life of the tubes and the optimal furnace retubing time must be accurately determined.

From the analyses of the manufacturer's failure data, they observed the following:

(a) Given a fixed tube temperature and tube pressure, the life distribution of the tube is known.

(b) The average life or any life percentile of the tube as a function of tube temperature and tube pressure is obtained.

Probability p that a tube will survive beyond t hours is obtained as follows:

$$p = 1 - q = \begin{cases} 1 - \Phi(\frac{\log t - \log t_{50}}{\sigma_0}), & \text{for log normal,} \\ e^{-(\frac{t}{t_{63.2}})^\beta}, & \text{for Weibull,} \end{cases}$$

where $\Phi(x) = \int_{-\infty}^{x} \frac{1}{\sqrt{2\pi}} e^{-x^2/2} dx$, i.e., a standard normal distribution function. Other parameters will be explained as follows.

The life percentile of the tube as a function of tube temperature and tube pressure can be expressed as:

$$\log t_p = C_0 + C_1 \frac{1}{T} + C_2 \frac{\log \sigma}{T} + C_3 (\frac{\log \sigma}{T})^2 + C_4 (\frac{\log \sigma}{T})^3,$$

where T is the tube metal temperature in Kelvin (T=273+the measured tube metal temperature in Celsius), t_p is the pth percentile of tube life ($\Pr\{Y \leq t_p\} = p/100$), σ is tube pressure in kilopounds per square inch, and C_0, C_1, C_2, C_3, C_4 are constants obtained from regression analysis of the manufacturer's tube failure data. The parameters σ_0 and β can be obtained from the data. σ_0 is log normal distribution's shape parameter, that is, $\log Y$ has normal distribution with mean $\log t_{50}$ and standard deviation σ_0.

Once the remaining life of each individual tube is estimated, that is, p is obtained, how do we estimate the remaining life of the system of the tubes? To answer this question, first the system failure should be defined. As shown in the above figure, the system failure is defined as follows: the furnace fails if at least 10 consecutive tubes all failed in a row, or at least one 2×2 grid of tubes all failed. The k-out-of-n system and consecutive-k-out-of-n system can be used for the remaining life estimation of the furnace.

In their paper, they just discussed the single failure criterion for system by using Barlow, Heidtmann and Rushdi's algorithm,

$$R(i,j) = q_j R(i,j-1) + p_j R(i-1,j-1),$$

$$F(i,j) = q_j F(i,j-1) + p_j R(i-1,j-1)$$

for the k-out-of-n system ($R(i,j)$ and $F(i,j)$ indicate the reliability and unreliability of an i-out-of-j:G system, respectively), using Hwang's efficient algorithm,

$$Q_L(n;k) = Q_L(n-1;k) + R_L(n-k-1;k)p_{n-k} \prod_{j=n-k+1}^{n} q_j$$

for consecutive-k-out-of-n:F system ($p_0 \equiv 1, Q_L(i,j)$ and $R_L(i;j)$ are the unreliability and reliability of a consecutive-i-out-of-j:F system, respectively), using Salvia and Lasher's two-dimensional consecutive-k-out-of-n:F system concept.

But as shown in the (n, f, k) example, we need to combine these failure criteria to form the system reliability. Of course, in this case, those combinations are not easy, specially, when a two-dimensional consecutive system is involved. Zuo and Wu [272] gave approximation results.

Last, we want to mention that Zuo and Wu [272] developed an optimization model for decision making regarding the retubing of the furnace system.

References

[1] M. A. Agin and A. P. Godbole, A new exact runs test for randomness, *Computing Science and Statistics, Proc. of the 22nd Symposium on the Interface* (C. Page and R. LePage eds.), Springer-Verlag, New York (1992) 281-285.

[2] S. Aki, Distributions of runs and consecutive systems on desired trees, *Ann. Inst. Stat. Math.* 51 (1999) 1-15.

[3] S. Aki and K. Hirano, Lifetime distribution and estimation problems of consecutive-k-out-of-n:F systems, *Ann. Inst. Stat. Math.* 48 (1996) 185-199.

[4] S. Aki and K. Hirano, Lifetime distributions of consecutive-k-out-of-n:F systems, *Nonlinear Analysis, Theor. Meth. Applica.* 30 (1997) 555-562.

[5] P. D. Alevizos, S. G. Papstavridis and P. Sypsas, Reliability of cyclic m-consecutive-k-out-of-n:F system, reliability, quality control and risk assessment, *Proc. IASTED Conference* (1992).

[6] N. Alon and J. Spencer, *The Probabilistic Method*, Wiley, New York, 1992.

[7] I. Antonopoulou and S. Papstavridis, Fast recursive algorithm to evaluate the reliability of a circular consecutive-k-out-of-n:F system, *IEEE Trans. Reliab.* R-36 (1987) 83-84.

[8] T. M. Apostol, *Mathematical Analysis*, 2nd edition, Addition-Wesley, Reading, Mass., 1974.

[9] T. M. Apostol, Binary vectors with prescribed subsets of consecutive ones, *J. Combin. Theory, Series A* 47 (1988) 176-190.

[10] C. Asano, Runs test for a circular distribution and a table of probabilities, *Ann. Inst. Stat. Math.* 17 (1965) 331-346.

[11] A. D. Barbour, O. Chryssaphinou and M. Roos, Compound Poisson approximation in reliability theory, *IEEE Trans. Reliab.* 44 (1995) 398-402.

[12] A. D. Barbour, O. Chryssaphinou and M. Roos, Compound Poisson approximation in systems reliability, *Nav. Research Log.* 43 (1996) 251-264.

[13] A. D. Barbour and G. K. Eagleson, Poisson convergence for dissociated statistics, *J. R. Statist. Soc.* B 46 (1984) 397-402.

[14] A. D. Barbour and L. Holst, Some applications of the Stein-Chen method for proving Poisson convergence, *Adv. Appl. Prob.* 21 (1989) 74-90.

[15] A. D. Barbour, L. Holst and S. Janson, *Poisson Approximation*, Oxford University Press, New York, 1992.

[16] R. E. Barlow and F. Proschan, *Statistical Theory of Reliability and Life Testing*, Holt, Reinhart and Winston, New York, 1975.

[17] D. E. Barton and F. N. David, Runs in a ring, *Biometrika* 45 (1958) 572-578.

[18] G. Bateman, On the power function of the longest run as a test for randomness in a sequence of alternatives, *Biometrika* 35 (1948) 97-112.

[19] A. Behr and L. Camarinopoulos, Two formulas for computing the reliability of incomplete k-out-of-n:G systems, *IEEE Trans. Reliab.* 46 (1997) 421-429.

[20] L. A. Belfore II, An $O((n\cdot(\log_2(n))^2)$ algorithm for computing the reliability of k-out-of-n:G and k-to-l-out-of-n:G systems, *IEEE Trans. Reliab.* 44 (1995) 132-136.

[21] J. Biernat, A unified algorithm of reliability estimation of k-out-of-n systems implied by the structure-function, *Microelectron. Reliab.* 33 (1993) 395-401.

[22] M. T. R. Blackwell, The effect of short production runs on CSP-1, *Technometrics* 19 (1977) 259-263.

[23] T. K. Boehme, A. Kossow and W. Preuss, A generalization of consecutive-k-out-of-n:F systems, *IEEE Trans. Reliab.* 41 (1992) 451-457.

[24] P. J. Boland and S. Papastavridis, Consecutive k out of n: F systems with cycle k, *Stat. Prob. Letters* 44 (1999) 155-160.

[25] P. J. Boland, F. Proschan and Y. L. Tong, Optimal arrangement of components via pairwise rearrangements, *Nav. Research Log.* 36 (1989) 807-815.

[26] P. J. Boland, F. Proschan and Y. L. Tong, Linear dependence in consecutive k-out-of-n:F system, *Prob. Eng. Inform. Sciences* 4 (1990) 391-397.

[27] R. C. Bollinger, Direct computation for consecutive-k-out-of-n:F systems, *IEEE Trans. Reliab.* R-31 (1982) 444-446.

[28] R. C. Bollinger, Reliability and runs of ones, *Math. Magazine* 57 (1984) 34-37.

[29] R. C. Bollinger, Strict consecutive-k-out-of-n:F systems, *IEEE Trans. Reliab.* R-34 (1985) 50-52.

[30] R. C. Bollinger and A. A. Salvia, Consecutive-k-out-of-n:F networks, *IEEE Trans. Reliab.* R-31 (1982) 53-56.

[31] R. C. Bollinger and A. A. Salvia, Consecutive-k-out-of-n:F system with sequential failures, *IEEE Trans. Reliab.* R-34 (1985) 43-45.

[32] E. J. Burr and G. Cane, Longest run of consecutive observations having a specified attribute, *Biometrika* 48 (1961) 461-465.

[33] J. Cai, Reliability of a large consecutive-k-out-of-r-from-n:F system with unequal component-reliability, *IEEE Trans. Reliab.* 43 (1994) 107-111.

[34] E. R. Canfield and W. P. McCormick, Asymptotic reliability of consecutive-k-out-of-n systems, *J. Appl. Prob.* 29 (1992) 142-155.

[35] F. R. Castella, Sliding window detection probabilities, *IEEE Trans. Aerosp. Electron. Systems* AES-12 (1976) 815-819.

[36] S. Chadjiconstantinidis and M. V. Koutras, Measures of component importance for Markov chain imbeddable reliability structures, *Nav. Research Log.* 46 (1999) 613-639.

[37] F.-Y. Chan, L. K. Chan and G. D. Lin, On consecutive-k-out-of-n:F systems, *Europ. J. Oper. Research* 36 (1988) 207-216.

[38] G. J. Chang, L. Cui and F. K. Hwang, Reliability for (n, f, k) systems, *Stat. Prob. Letters* 43 (1999) 237-242.

[39] G. J. Chang, L. Cui and F. K. Hwang, New comparisons in Birnbaum importance for the consecutive-k-out-of-n system, *Prob. Eng. Inform. Sciences* 13 (1999) 187-192.

[40] G. J. Chang and F. K. Hwang, Optimal consecutive-k-out-of-n systems under a fixed budget, *Prob. Eng. Inform. Sciences* 2 (1988) 63-73.

[41] G. J. Chang, F. K. Hwang and Cui, Corrigenda on "New comparisons in Birnbaum importance for the consecutive-k-out-of-n system", *Prob. Eng. Inform. Sciences* 14 (2000) 405.

[42] H.-W. Chang, *Optimal Assignment of Consecutive-k Systems*, Ph.D Thesis, Dept. Comput. Sci. and Inform. Eng., National Chiao Tung Univ., June 2000.

[43] H.-W. Chang, R. J. Chen and F. K. Hwang, The structural Birnbaum importance of consecutive-k system, *J. Combin. Opt.* to appear.

[44] H.-W. Chang and F. K. Hwang, Existence of invariant series consecutive-k:G systems, *IEEE Trans. Reliab.* 48 (1999) 306-308.

[45] H.-W. Chang and F. K. Hwang, Rare event component importance of the consecutive-k system, preprint (2000).

[46] J. C. Chang, R. J. Chen and F. K. Hwang, A fast reliability algorithm for the circular consecutive-weighted-k-out-of-n:F system, *IEEE Trans. Reliab.* 47 (1998) 472-474.

[47] J. C. Chang, R. J. Chen and F. K. Hwang, Faster algorithms to compute the reliabilities of the (n, f, k) systems, preprint (2000).

[48] M.-T. Chao, The finite time behavior of CSP when defects are dependent, *Proc. Nat. Sci. Council ROC(A)* 13 (1989) 18-22.

[49] M. T. Chao and J. C. Fu, A limit theorem of certain repairable systems, *Ann. Inst. Stat. Math.* 41 (1989) 809-818.

[50] M. T. Chao and J. C. Fu, The reliability of a large series system under Markov structure, *Adv. Appl. Prob.* 23 (1991) 894-908.

[51] M. T. Chao, J. C. Fu and M. V. Koutras, Survey of reliability studies of consecutive-k-out-of-n:F and related systems, *IEEE Trans. Reliab.* 44 (1995) 120-127.

[52] M. T. Chao and G. D. Lin, Economical design of large consecutive-k-out-of-n:F systems, *IEEE Trans. Reliab.* R-33 (1984) 411-413.

[53] C. A. Charalambides, Lucas numbers and polynomials of order k and the length of the longest circular success run, *Fibonacci Quarterly* 29 (1991) 290-297.

[54] L. H. Y. Chen, Poisson approximation for dependent trials, *Ann. Prob.* 3 (1975) 534-545.

[55] R. W. Chen and F. K. Hwang, Failure distributions of consecutive-k-out-of-n:F systems, *IEEE Trans. Reliab.* R-34 (1985) 338-341.

[56] R. W. Chen, F. K. Hwang and W.-C. W. Li, Consecutive-2-out-of-n:F systems with node and link failures, *IEEE Trans. Reliab.* 42 (1993) 497-502.

[57] R. W. Chen, F. K. Hwang and W.-C. W. Li, A reversible model for consecutive-2-out-of-n:F systems with node and link failures, *Prob. Eng. Inform. Sciences* 8 (1994) 189-200.

[58] C.-C. Chen and K.-M. Koh, *Principle and Techniques in Combinatorics*, World Scientific, Singapore, 1992, pp. 44 and 169.

[59] C. H. Cheng, Fuzzy consecutive-k-out-of-n:F system reliability, *Microelectron. Reliab.* 34 (1994) 1909-1922.

[60] D. T. Chiang and R.-f. Chiang, Relayed communication via consecutive-k-out-of-n:F system, *IEEE Trans. Reliab.* R-35 (1986) 65-67.

[61] D. T. Chiang and S.-C. Niu, Reliability of consecutive-k-out-of-n:F system, *IEEE Trans. Reliab.* R-30 (1981) 87-89.

[62] O. Chryssaphinou and S. Papastavridis, Reliability of a consecutive-k-out-of-n system in a random environment, *J. Appl. Prob.* 27 (1990) 452-458.

[63] O. Chryssaphinou and S. G. Papastavridis, Limit distribution for a consecutive-k-out-of-n:F system, *Adv. Appl. Prob.* 22 (1990) 491-493.

[64] O. Chryssaphinou, S. Papastavridis and P. Sypsas, Relayed communication via parallel redundancy, *IEEE Trans. Reliab.* 38 (1989) 454-456.

[65] L. Cui, A. G. Hawkes and A. Jalali, The increasing failure rate property of consecutive-k-out-of-n systems, *Prob. Eng. Inform. Sci.* (1995) 217-225.

[66] H. M. Dao and C. B. Silio, Jr., Ring-network with a constrained number of consecutively-bypassed stations, *IEEE Trans. Reliab.* 47 (1998) 35-43.

[67] J. L. Denny and S. J. Yakowitz, Admissible run-contingency type tests for independence and Markov dependence, *J. Amer. Stat. Assoc.* 73 (1978) 177-181.

[68] C. Derman, G. J. Lieberman and S. M. Ross, On the consecutive-k-out-of-n:F system, *IEEE Trans. Reliab.* R-31 (1982) 57-63.

[69] D. Z. Du and F. K. Hwang, Optimal consecutive-2 systems of lines and cycles, *Networks* 15 (1985) 439-447.

[70] D. Z. Du and F. K. Hwang, Optimal consecutive-2-out-of-n systems, *Math. Oper. Research* 11 (1986) 187-191.

[71] D. Z. Du and F. K. Hwang, Optimal assignments for consecutive-2 graphs, *SIAM Alg. Disc. Meth.* 8 (1987) 510-518.

[72] D. Z. Du and F. K. Hwang, Reliabilities of consecutive-2 graphs, *Prob. Eng. Inform. Sciences* 1 (1987) 293-298.

[73] D. Z. Du and F. K. Hwang, A direct algorithm for computing reliability of a consecutive-k cycle, *IEEE Trans. Reliab.* 37 (1988) 70-72.

[74] S. S. Elias, N. A. Mokhles and E. M. Elsayed, Consecutive k-out-of-n:F system with sequential failures and a single repair, *Microelectron. Reliab.* 34 (1994) 39-51.

[75] W. Feller, *An Introduction to Probability Theory and Its Applications*, Volume 1, 3rd ed., Wiley, New York (1968).

[76] D. E. Ferguson, An expression for generalized Fibonacci numbers, *Fibonacci Quart.* 4 (1966) 270-273.

[77] J. C. Fu, Reliability of a large consecutive-k-out-of-n:F system, *IEEE Trans. Reliab.* R-34 (1985) 127-130.

[78] J. C. Fu, Bounds for reliability of large consecutive-k-out-of-n:F systems with unequal component reliability, *IEEE Trans. Reliab.* R-35 (1986) 316-319.

[79] J. C. Fu, Reliability of consecutive-k-out-of-n:F systems with ($k-1$)-step Markov dependence, *IEEE Trans. Reliab.* R-35 (1986) 602-606.

[80] J. C. Fu, Poisson convergence in reliability of a large linearly connected system as related to coin tossing, *Statistica Sinica* 3 (1993) 261-275.

[81] J. C. Fu and B. Hu, On reliability of a large consecutive-k-out-of-n:F system with ($k-1$)-step Markov dependence, *IEEE Trans. Reliab.* R-36 (1987) 75-77.

[82] J. C. Fu and M. V. Koutras, Poisson approximations for 2-dimensional patterns, *Ann. Inst. Stat. Math.* 46 (1994) 179-192.

[83] G. Ge and L. Wang, Exact reliability formula for consecutive-k-out-of-n:F systems with homogeneous Markov dependence, *IEEE Trans. Reliab.* 39 (1990) 600-602.

[84] A. P. Godbole, Specific formula for some success run distributions, *Stat. Prob. Letters* 10 (1990) 119-124.

[85] A. P. Godbole, Degenerate and Poisson convergence criteria for success runs, *Stat. Prob. Letters* 10 (1990) 247-255.

[86] A. P. Godbole, On hypergeometric and related distributions of order k, *Commun. Stat.-Theory Meth.* 19 (1990) 1291-1301.

[87] A. P. Godbole, Poisson approximations for runs and patterns of rare events, *Adv. Appl. Prob.* 23 (1991) 851-865.

[88] A. P. Godbole, Approximate reliabilities of m-consecutive-k-out-of-n: failure systems, *Statistica Sinica* 3 (1993) 321-327.

[89] A. P. Godbole, L. K. Potter and J. K. Sklar, Improved upper bounds for the reliability of d-dimensional consecutive-k-out-of-n:F systems, *Nav. Research Log.* 45 (1998) 219-230.

[90] S. W. Golomb, Distributions of periods and frequencies of runs in random binary sequences, *Adv. Appl. Math.* 10 (1989) 278-284.

[91] H. W. Gould, *Combinatorial Identities*, revised edition, Morgantown, WV, 1972.

[92] I. P. Goulden, Generating functions and reliabilities for consecutive-k-out-of-n systems, *Utilitas Math.* 32 (1987) 141-147.

[93] R. L. Graham, D. E. Knuth and O. Patashnik, *Concrete Mathematics*, Addision-Wesley, New York (1994).

[94] I. Greenberg, The first occurrence of n successes in N trails, *Technometrics* 12 (1970) 627-634.

[95] W. S. Griffith, On consecutive k-out-of-n failure systems and their generalizations, *Reliability and Quality Control* (A. P. Basu ed.), Elsevier Science Publ. B.V., Amsterdam (1986) 157-165.

[96] W. S. Griffith and Z. Govindarajulu, Consecutive-k-out-of-n failure systems: reliability, availability, component importance and multistate extensions, *Amer. J. Math. Manag. Sci.* 5 (1985) 125-160.

[97] H. Harborth, Endliche 0-1-folgen mit gleichen teibloechen, *J. Reine Angewandte Math.* (1974) 139-154.

[98] Y. Higashiyama, H. Ariyoshi and M. Kraetzl, Fast solutions for consecutive-2-out-of-r-from-n:F system, *IEICE Trans. Fundament. Electron. Commun. Compt. Sci.* E78A (1995) 680-684.

[99] K. Hirano, Some properties of the distributions of order k, *Fibonacci Numbers and Their Applications* (A. N. Philippou, G. E. Bergum and A. F. Horadam eds.), D. Reidel, Dordrecht (1986) 43-53.

[100] R. Huntington, Mean recurrence times for k success within m trails, *J. Appl. Prob.* 13 (1976) 604-607.

[101] F. K. Hwang, Fast solutions for consecutive-k-out-of-n:F system, *IEEE Trans. Reliab.* R-31 (1982) 447-448.

[102] F. K. Hwang, Selecting k objects from a cycle with p pairs of separation s, *J. Combin. Theory, Series A* 37 (1984) 197-199.

[103] F. K. Hwang, Simplified reliabilities for consecutive-k-out-of-n systems, *SIAM J. Alg. Disc. Meth.* 7 (1986) 258-264.

[104] F. K. Hwang, Relayed consecutive-k-out-of-n:F lines, *IEEE Trans. Reliab.* 37 (1988) 512-514.

[105] F. K. Hwang, Invariant permutations for consecutive k-out-of-n cycles, *IEEE Trans. Reliab.* 38 (1989) 65-67.

[106] F. K. Hwang, Optimal assignment of components to a two-stage k-out-of-n system, *Math. O. R.* 14 (1989) 376-382.

[107] F. K. Hwang, An explicit solution for the number of minimal p-cutsequences in a consecutive-k-out-of-n:F system, *IEEE Trans. Reliab.* 40 (1991) 553-554.

[108] F. K. Hwang, An $O(kn)$-time algorithm for computing the reliability of a circular consecutive-k-out-of-n:F system, *IEEE Trans. Reliab.* 42 (1993) 161-162.

[109] F. K. Hwang, A new index for component importance, *Oper. Res. Lett.*, to appear.

[110] F. K. Hwang, L. R. Cui, J. C. Chang and W. D. Lin, Comments on "Reliability and component importance of a consecutive-k-out-of-n system" by Zuo, *Microelectron. Reliab.* 40 (2000) 1061-1063.

[111] F. K. Hwang and W.-C. W. Li, Connectivity reliabilities and Hamiltonian reliabilities of linear and circular consecutive-2 link systems, *Inter. J. Reliab. Quality Safety Eng.* 1 (1994) 247-256.

[112] F. K. Hwang and C. K. Pai, Sequential construction of a consecutive-2 system, *Inform. Proc. Lett.*, to appear.

[113] F. K. Hwang and S. G. Papastavridis, Binary vectors with exactly k nonoverlapping m-tuples of consecutive ones, *Discrete Appl. Math.* 30 (1991) 83-86. Erratum, the same journal 46 (1993) 283.

[114] F. K. Hwang and U. G. Rothblum, Optimality of monotone assemblies for coherent systems composed of series modules, *Oper. Research* 42 (1994) 709-720.

[115] F. K. Hwang and D. Shi, Redundant consecutive-k systems, *Oper. Research Letters* 6 (1987) 293-296.

[116] F. K. Hwang and D. Shi, Optimal relayed mobile communication systems, *IEEE Trans. Reliab.* 38 (1989) 457-459.

[117] F. K. Hwang and P. E. Wright, An $O(k^3 \cdot \log(n/k))$ algorithm for the consecutive-k-out-of-n:F system, *IEEE Trans. Reliab.* 44 (1995) 128-131

[118] F. K. Hwang and P. E. Wright, An $O(n \log n)$ algorithm for the generalized birthday problem, *Comput. Stat. Data Analysis* 23 (1997) 443-451.

[119] F. K. Hwang and Y. C. Yao, Multistate consecutively-connected systems, *IEEE Trans. Reliab.* 38 (1989) 472-474.

[120] F. K. Hwang and Y. C. Yao, On the failure rates of consecutive-k-out-of-n systems, *Prob. Eng. Inform. Sciences* 4 (1990) 57-71.

[121] F. K. Hwang and Y. C. Yao, A direct argument for Kaplansky's theorem on cyclic arrangement and its generalization, *Oper. Research Letters* 10 (1991) 241-143.

[122] S. Iyer, Distribution of time to failure of consecutive k-out-of-n:F systems, *IEEE Trans. Reliab.* 39 (1990) 97-101.

[123] A. Jalali, A. G. Hawkes, L. Cui and F. K. Hwang, The optimal consecutive-k-out-of-n:G line for $n \leq 2k$, preprint (1999).

[124] C. Jennen-Steinmetz and T. Gasser, The asymptotic power of runs tests, *Scand. J. Stat.* 13 (1986) 263-269.

[125] R. Jiang, M. J. Zuo and H. X. Li, Weibull and inverse Weibull mixture models allowing negative weights, *Reliab. Eng. Syst. Safety* 66 (1999) 227-234.

[126] K.-H. Jung and H. Y. Kim, Linear consecutive-k-out-of-n:F system reliability with common-mode forced outages, *Reliab. Eng. Syst. Safety* 41 (1993) 49-55.

[127] S. C. Kao, Computing reliability from Warranty, *Proc. Amer. Stat. Assoc. 142nd Meeting, Section on Stat. Comp.* (1982) 309-312.

[128] J. Kaplansky, Solution of the "problème des ménages", *Bull. Amer. Math. Soc.* 49 (1943) 784-785.

[129] S. Karlin and J. McGregor, Coincidence probabilities, *Pacific J. Math.* 9 (1959) 1141-1164.

[130] J. M. Kontoleon, Analysis of a dynamic redundant system, *IEEE Trans. Reliab.* R-27 (1978) 116-119.

[131] J. M. Kontoleon, Reliability determination of a r-successive-out-of-n:F system, *IEEE Trans. Reliab.* R-29 (1980) 437.

[132] J. Konvalina, On the number of combinations without unit separation, *J. Combin. Theory, Series A* 31 (1981) 101-107.

[133] A. Kossow and W. Preuss, Failure probability of strict consecutive-k-out-of-n:F systems, *IEEE Trans. Reliab.* R-36 (1981) 551-553.

[134] A. Kossow and W. Preuss, Reliability of consecutive-k-out-of-n:F systems with nonidentical component reliabilities, *IEEE Trans. Reliab.* 38 (1989) 229-233.

[135] A. Kossow and W. Preuss, Mean time-to-failure of a linear-consecutive-k-out-of-n:F system, *IEEE Trans. Reliab.* 40 (1991) 271-272.

[136] A. Kossow and W. Preuss, Reliability of linear consecutively-connected systems with multistate components, *IEEE Trans. Reliab.* 44 (1995) 518-522.

[137] S. Kounias and M. Sfakianakis, The reliability of a linear system and its connection with the generalized birthday problem, *Statistica Applicata* 3 (1991) 531-543.

[138] S. Kounias and M. Sfakianakis, A combinatorial problem associated with the reliability of a circular system, *Proc. Hermis 1992 Conference* (1992).

[139] S. Kounias and K. Sotirakoglou, Bonferroni bounds revised, *J. Appl. Prob.* 26 (1989) 233-241.

[140] M. V. Koutras, On a Markov chain approach for the study of reliability structures, *J. Appl. Prob.* 33 (1996) 357-367.

[141] M. V. Koutras, Consecutive-k, r-out-of-n:DFM systems, *Microelectron. Reliab.* 37 (1997) 597-603.

[142] M. V. Koutras and S. G. Papastavridis, Application of the Stein-Chen method for bounds and limit theorems in the reliability of coherent structures, *Nav. Research Log.* 40 (1993) 617-631.

[143] M. V. Koutras and S. G. Papastavridis, On the number of runs and related statistics, *Statistica Sinica* 3 (1993) 277-294.

[144] M. V. Koutras, G. K. Papadopoulos and S. G. Papastavridis, Reliability of 2-dimensional consecutive-k-out-of-n:F systems, *IEEE Trans. Reliab.* 42 (1993) 658-661.

[145] M. V. Koutras, G. K. Papadopoulos and S. G. Papastavridis, A reliability bound for 2-dimensional consecutive-k-out-of-n:F systems, *Nonlinear Analysis, Theor. Meth. Applica.* 30 (1997) 3345-3348.

[146] B. Ksir, Comment on: 2-dimensional consecutive-k-out-of-n:F models, *IEEE Trans. Reliab.* 41 (1992) 575.

[147] B. Ksir and M. Boushaba, Reliability bounds and direct computation of the reliability of a consecutive k-out-of-n:F system with Markov dependence, *Microelectron. Reliab.* 33 (1993) 313-317.

[148] U. D. Kumar and M. N. Gopalan, Analysis of consecutive k-out-of-n:F systems with single repair facility, *Microelectron. Reliab.* 37 (1997) 587-590.

[149] W. Kuo, J. C. Fu, W. Y. Lou and F. K. Hwang, Opinions on consecutive-k-out-of-n:F systems, *IEEE Trans. Reliab.* 43 (1994) 659-662.

[150] W. Kuo, W. Zhang and M. Zuo, A consecutive-k-out-of-n:G system: the mirror image of consecutive-k-out-of-n:F system, *IEEE Trans. Reliab.* 39 (1990) 244-253.

[151] Y. Lam and Y. L. Zhang, Repairable consecutive-k-out-of-n:F system with Markov dependence, *Nav. Research Log.* 47 (2000) 18-39.

[152] M. Lambiris and S. Papastavridis, Exact reliability formulas for linear and circular consecutive-k-out-of-n:F systems, *IEEE Trans. Reliab.* R-34 (1985) 124-126.

[153] R. J. Larsen, C. L. Holmes and C. W. Heath, Jr., A statistical test for measuring unimodal clustering: a description of the test and its application to cases of acute leukemia in Metropolitan Atlanta, Georgia, *Biometrics* 29 (1973) 301-309.

[154] A. Lascoux, Suites récurrentes linéaires, *Adv. Appl. Math.* 7 (1986) 228-235.

[155] T. S. Lau, The reliability of exchangeable binary systems, *Stat. Prob. Letters* 13 (1992) 153-158.

[156] J. Liao, A generalized formula for computing the reliability of consecutive-k-out-of-n:F circular systems, *Kexue Tongbao* 31 (1986) 1533-1537.

[157] F. H. Lin, W. Kuo and F. Hwang, Structure importance of consecutive-k-out-of-n systems, *Oper. Research Letters* 25 (1999) 101-107.

[158] F. S. Makri and A. N. Philippou, Binomial distributions of order k on the circle, *Runs and Patterns: Selected Papers* (A. Godbole, ed.) *Appl. Math.* 283 (1993) 65-81.

[159] F. S. Makri and A. N. Philippou, Exact reliability formulas for linear and circular m-consecutive-k-out-of-n:F systems, *Microelectron. Reliab.* 36 (1996) 657-660.

[160] F. S. Makri and Z. M. Psillakis, Bounds for reliability of k-within two-dimensional consecutive-r-out-of-n failure systems, *Microelectron. Reliab.* 36 (1996) 341-345.

[161] F. S. Makri and Z. M. Psillakis, Bounds for reliability of k-within connected-(r,s)-out-of-(m,n) failure systems, *Microelectron. Reliab.* 37 (1997) 1217-1224.

[162] J. Malinowski and W. Preuss, Reliability of circular consecutively-connected systems with multistate components, *IEEE Trans. Reliab.* 44 (1995) 532-534.

[163] J. Malinowski and W. Preuss, A recursive algorithm evaluating the exact reliability of a consecutive k-within-m-out-of-n:F system, *Microelectron. Reliab.* 35 (1995) 1461-1465.

[164] J. Malinowski and W. Preuss, Lower and upper bounds for the reliability of consecutive-(r,s)-out-of-(m,n):F lattice systems, *IEEE Trans. Reliab.* 45 (1996) 156-160.

[165] J. Malinowski and W. Preuss, Reliability of reverse-tree-structured systems with multistate components, *Microelectron. Reliab.* 36 (1996) 1-7.

[166] J. Malinowski and W. Preuss, Reliability evaluation for tree-structured systems with multistate components, *Microelectron. Reliab.* 36 (1996) 9-17.

[167] J. Malinowski and W. Preuss, A recursive algorithm evaluating the exact reliability of a circular consecutive k-within-m-out-of-n:F system, *Microelectron. Reliab.* 36 (1996) 1389-1394.

[168] J. Malinowski and W. Preuss, Reliability increase of consecutive-k-out-of-n:F and related systems through components' rearrangement, *Microelectron. Reliab.* 36 (1996) 1417-1423.

[169] J. Malinowski and W. Preuss, Reliability of a two-way linear consecutively connected system with multistate components, *Microelectron. Reliab.* 36 (1996) 1483-1488.

[170] J. Malinowski and W. Preuss, Reliability of a two-way circular consecutively connected system with multistate components, *Microelectron. Reliab.* 37 (1997) 1255-1258.

[171] D. M. Malon, Optimal consecutive-2-out-of-n:F component sequencing, *IEEE Trans. Reliab.* R-33 (1984) 414-418.

[172] D. M. Malon, Optimal consecutive-k-out-of-n:F component sequencing, *IEEE Trans. Reliab.* R-34 (1985) 46-49.

[173] A. de Moivre, The doctrine of chance, *Chelsea Publishing Co.*, 3rd Edition (1967) 254-259.

[174] A. M. Mood, The distribution theory of runs, *Math. Stat.* 11 (1940) 367-392.

[175] W. O. J. Moser and M. Abramson, Enumeration of combinations with restricted differences and cospan, *J. Comb. Theory* 7 (1969) 162-170.

[176] F. Mosteller, Note on an application of runs to quality control charts, *Ann. Math. Stat.* 12 (1941) 228-232.

[177] J. Naus, An extension of the birthday problem, *The Amer. Statistician* 22 (1968) 27-29.

[178] J. Naus, Probabilities for a generalized birthday problem, *J. Amer. Stat. Assoc.* 69 (1974) 810-815.

[179] J. B. Nelson, Minimal-order models for false-alarm calculations on sliding windows, *IEEE Trans. Aerospace Electronic Systems* AES-14 (1978) 351-363.

[180] K. Noguchi, M. Sasaki, S. Yanagi and T. Yuge, Reliability of connected-(r,s)-out-of-(m,n):F system, *Electron. Commun. Japan (Part 3)* 80 (1997) 27-35.

[181] K. Noguchi, M. Sasaki, S. Yanagi and T. Yuge, System reliability of a sensor system considering the detectable area, *Electron. Commun. Japan (Part 3)* 80 (1997) 64-72.

[182] P. C. O'Brien, A test for randomness, *Biometrics* 32 (1976) 391-401.

[183] J. Panaretos and E. Xekalaki, On some distributions arising from certain generalized sampling schemes, *Commun. Stat.-Theory Meth.* 15 (1986) 873-891.

[184] J. Panaretos and E. Xekalaki, A probability distribution associated with events with multiple occurrences, *Stat. Prob. Letters* 8 (1989) 389-395.

[185] S. Papastavridis, Upper and lower bounds for the reliability of consecutive-k-out-of-n:F system, *IEEE Trans. Reliab.* R-35 (1986) 607-610.

[186] S. G. Papastavridis, Algorithms for strict consecutive-k-out-of-n:F system, *IEEE Trans. Reliab.* R-35 (1986) 613-615.

[187] S. Papastavridis, A limit theorem for the reliability of a consecutive-k-out-of-n system, *Adv. Appl. Prob.* 19 (1987) 746-748.

[188] S. Papastavridis, The most important component in a consecutive-k-out-of-n:F system, *IEEE Trans. Reliab.* R-36 (1987) 266-268.

[189] S. G. Papastavridis, A Weibull limit for the reliability of a consecutive k-within-m-out-of-n system, *Adv. Appl. Prob.* 20 (1988) 690-692.

[190] S. Papastavridis, The number of failed components in a consecutive-k-out-of-n:F system, *IEEE Trans. Reliab.* 38 (1989) 338-340.

[191] S. G. Papastavridis, Lifetime distribution of circular consecutive-k-out-of-n:F systems with exchangeable lifetimes, *IEEE Trans. Reliab.* 38 (1989) 460-461.

[192] S. Papastavridis, m-Consecutive-k-out-of-n:F systems, *IEEE Trans. Reliab.* 39 (1990) 386-388.

[193] S. G. Papastavridis, Comment on: strict consecutive-k-out-of-n:F systems, *IEEE Trans. Reliab.* 42 (1993) 171.

[194] S. G. Papastavridis and O. Chrysaphinou, An approximation for large consecutive-k-out-of-n:F systems, *IEEE Trans. Reliab.* 37 (1988) 386-387.

[195] S. Papastavridis and J. Hadjichristos, Mean time to failure for a consecutive-k-out-of-n:F system, *IEEE Trans. Reliab.* R-36 (1987) 85-86.

[196] S. Papastavridis and I. Hadzichristos, Formulas for the reliability of a consecutive-k-out-of-n:F system, *J. Appl. Prob.* 26 (1988) 772-779.

[197] S. G. Papastavridis and M. V. Koutras, Consecutive k-out-of-n systems with maintenance, *Ann. Inst. Stat. Math.* 44 (1992) 605-612.

[198] S. G. Papastavridis and M. V. Koutras, Bounds for reliability of consecutive k-within-m-out-of-n:F systems, *IEEE Trans. Reliab.* 42 (1993) 156-160.

[199] S. Papastavridis and M. Lambiris, Reliability of a consecutive-k-out-of-n:F system for Markov-dependent components, *IEEE Trans. Reliab.* R-36 (1987) 78-79.

[200] S. G. Papastavridis and M. E. Sfakianakis, Optimal-arrangement and importance of the components in a consecutive-k-out-of-n:F system, *IEEE Trans. Reliab.* 40 (1991) 277-279.

[201] E. A. Pekoz and S. M. Ross, A simple derivation of exact reliability formulas for linear and circular consecutive-k-out-of-n:F systems, *J. Appl. Prob.* 32 (1995) 554-557.

[202] K. Petakos and T. Tsapelas, Reliability analysis for systems in a random environment, *J. Appl. Prob.* 34 (1997) 1021-1031.

[203] A. N. Philippou, Distributions and Fibonacci polynomials order k, longest runs, and reliability of consecutive-k-out-of-n:F systems, *Fibonacci Numbers and Their Applications* (Philippou et al, eds.), (1986) 203-227.

[204] A. N. Philippou and F. S. Makri, Longest success runs and Fibonacci-type polynomials, *Fibonacci Quarterly* 23 (1985) 338-346.

[205] A. N. Philippou and F. S. Makri, Success, runs and longest runs, *Stat. Prob. Letters* 4 (1986) 101-105.

[206] A. N. Philippou and F. S. Makri, Closed formulas for the failure probability of a strict-consecutive-k-out-of-n:F system, *IEEE Trans. Reliab.* R-36 (1987) 80-82.

[207] A. N. Philippou and F. S. Makri, Longest circular runs with an application in reliability via Fibonacci-type polynomials of order k, *Applications of Fibonacci Numbers* (G. E. Bergum et al. eds.), Kluwer Academic Pub. (1990) 281-286.

[208] R. R. Prairie and W. J. Zimmer and J. K. Brookhouse, Some acceptance sampling plans based on the theory of runs, *Technometrics* 4 (1962) 177-185.

[209] W. Preuss, On the reliability of generalized consecutive systems, *Nonlinear Analysis, Theor. Meth. Applica.* 30 (1997) 5425-5429.

[210] H. Prodinger, On the number of combinations without fixed distance, *J. Combin. Theory, Series A* 35 (1983) 362-365.

[211] J. S. Provan and M.O. Ball, The complexity of counting cuts and of computing the probability that a graph is connected, *SIAM J. Comput.* 12 (1983) 777-788.

[212] Z. M. Psillakis, A simulation algorithm for computing failure probability of a consecutive-k-out-of-r-from-n:F system, *IEEE Trans. Reliab.* 44 (1995) 523-531.

[213] L. Rade, A consecutive k-out-of-n reliability system in random environment, *Microelectron. Reliab.* 34 (1994) 1311-1318.

[214] A. M. Rushdi, A switching-algebraic analysis of consecutive-k-out-of-n:F systems, *Microelectron. Reliab.* 27 (1987) 171-174.

[215] A. M. Rushdi, A switching-algebraic analysis of circular consecutive-k-out-of-n:F systems, *Reliab. Eng. System Safety* 21 (1988) 119-127.

[216] A. M. Rushdi, On the analysis of ordinary and strict consecutive-k-out-of-n:F systems, *IEEE Trans. Reliab.* 37 (1988) 57-64.

[217] A. M. Rushdi, Effect of statistical dependencies in strict consecutive-k-out-of-n:F systems, *Microelectron. Reliab.* 28 (1988) 309-318.

[218] A. M. Rushdi, Some open questions on: strict consecutive-k-out-of-n:F systems, *IEEE Trans. Reliab.* 39 (1990) 380-381.

[219] A. A. Salvia, Simple inequalities for consecutive-k-out-of-n:F networks, *IEEE Trans. Reliab.* R-31 (1982) 450.

[220] A. A. Salvia, Author reply to "comment on: 2-dimensional consecutive-k-out-of-n:F models," *IEEE Trans. Reliab.* 41 (1992) 575.

[221] A. A. Salvia and W. C. Lasher, 2-Dimensional consecutive-k-out-of-n:F models, *IEEE Trans. Reliab.* 39 (1990) 382-385.

[222] M. Santha and Y. Zhang, Consecutive-2 systems on trees, *Prob. Eng. Inform. Sciences* 1 (1987) 441-456.

[223] B. Saperstein, The generalized birthday problem, *J. Amer. Stat. Assoc.* 67 (1972) 425-428.

[224] B. Saperstein, On the occurrences of n successes within N Bernoulli trials, *Technometrics* 15 (1973) 809-818.

[225] B. Saperstein, Note on a clustering problem, *J. Appl. Prob.* 12 (1975) 629-632.

[226] M. Sasaki, Correction to: mean time to failure for a consecutive-k-out-of-n:F system, *IEEE Trans. Reliab.* 41 (1992) 441.

[227] M. Sasaki, Y. Nakai and T. Yuge, Mean time to failure for a consecutive-k-out-of-n:F system, *IEEE Trans. Reliab.* 41 (1992) 306.

[228] M. Sasaki, N. Yamaguchi, T. Yuge and S. Yanagi, MTBF for consecutive-k-out-of-n:F systems with nonidentical component availabilities, *IEICE Trans. Fund. Electr. Comm. Comput. Sci.* E77A (1994) 122-128.

[229] M. Satam, Consecutive-k-out-of-n:F system: a comment, *IEEE Trans. Reliab.* 40 (1990) 62.

[230] R. J. Serfling, *Approximation Theorems of Mathematical Statistics*, Wiley, New York.

[231] M. Sfakianakis, Birnbaum importance of components in a consecutive-k-out-of-r-from-n:F system, Reliability Quality Control and Risk Assessment, *Proc. IASTED Conference* (1992) 36-37.

[232] M. Sfakianakis, Optimal arrangement of components in a consecutive k-out-of-r-from-n:F system, *Microelectron. Reliab.* 33 (1993) 1573-1578.

[233] M. Sfakianakis, S. Kounias and A. Hillaris, Reliability of a consecutive k-out-of-r-from-n:F system, *IEEE Trans. Reliab.* 41 (1992) 442-447.

[234] M. E. Sfakianakis and S. G. Papastavridis, Reliability of a general consecutive-k-out-of-n:F system, *IEEE Trans. Reliab.* 42 (1993) 491-496.

[235] J. G. Shanthikumar, Recursive algorithm to evaluate the reliability of a consecutive-k-out-of-n:F system, *IEEE Trans. Reliab.* R-31 (1982) 442-443.

[236] J. G. Shanthikumar, Lifetime distribution of consecutive-k-out-of-n:F systems with exchangeable lifetimes, *IEEE Trans. Reliab.* R-34 (1985) 480-483.

[237] J. G. Shanthikumar, Reliability of systems with consecutive minimal cutsets, *IEEE Trans. Reliab.* R-36 (1987) 546-549.

[238] J. G. Shanthikumar, Bounding network-reliability using consecutive minimal cutsets, *IEEE Trans. Reliab.* 37 (1988) 45-49.

[239] P. W. Shaughnessy, Multiple runs distributions: recurrences and critical values, *J. Amer. Stat. Assoc.* 76 (1981) 732-736.

[240] J. K. Shen and M. J. Zuo, A necessary condition for optimal consecutive k-out-of-n:G system design, *Micrelectron. Reliab.* 34 (1994) 485-493.

[241] J. Shen and M. J. Zuo, Optimal design of series consecutive-k-out-of-n:G systems, *Reliab. Eng. Syst. Safety* 45 (1994) 277-283.

[242] M. A. Stephens, Whitworth runs on a circle, *Ann. Inst. Stat. Math.* 29 (1977) 287-293.

[243] H. Sun and J. Liao, The reliability (n, f, k) system, *J. Electron.* 12 (1990) 436-439.

[244] Y. L. Tong, A rearrangement inequality for the longest run, with an application to network reliability, *J. Appl. Prob.* 22 (1985) 386-393.

[245] Y. L. Tong, Some new results on the reliability of circular consecutive-k-out-of-n:F system, *Reliability and Quality Control* (A. P. Basu, ed.), Elsevier Science Pub. (1986) 395-399.

[246] S. S. Tung, Combinatorial analysis in determining reliability, *Proc. Ann. Reliab. and Maintainability Symp.* (1982) 262-266.

[247] L. V. Utkin, A new heuristic method redundancy optimization in a consecutive-k-out-of-n:F system by fuzzy probability, *Microelectron. Reliab.* 33 (1993) 2197-2205.

[248] L. V. Utkin, Fuzzy reliability of repairable systems in the possibility context, *Microelectron. Reliab.* 34 (1994) 1865-1876.

[249] S. R. Wallenstein and I. J. Naus, Probabilities for a kth nearest neighbor problem on the line, *Annals Prob.* 1 (1973) 188-190.

[250] Y. H. Wang and S. X. Ji, Limit theorems for the number of occurrences of consecutive k successes in n Markovian trials, *J. Appl. Prob.* 32 (1995) 727-735.

[251] V. K. Wei, F. K. Hwang and V. T. Sös, Optimal sequencing of items in a consecutive-2-out-of-n system, *IEEE Trans. Reliab.* R-32 (1983) 30-33.

[252] J. Wolfowitz, On the theory of runs with some applications to quality control, *Ann. Math. Stat.* 14 (1943) 280-288.

[253] W. H. Woodall, The distribution of the run length of one-sided CUSUM procedures for continuous random variables, *Technometrics* 25 (1983) 295-301.

[254] W. H. Woodall, On the Markov chain approach to the two-sided CUSUM procedure, *Technometrics* 26 (1984) 41-46.

[255] J.-S. Wu and R.-J. Chen, An $O(kn)$ algorithm for a circular consecutive-k-out-of-n:F system, *IEEE Trans. Reliab.* 41 (1992) 303-305.

[256] J.-S. Wu and R.-J. Chen, Efficient algorithm for reliability of a circular consecutive-k-out-of-n:F system, *IEEE Trans. Reliab.* 42 (1993) 163-164.

[257] J.-S. Wu and R.-J. Chen, Efficient algorithms for k-out-of-n and consecutive-weighted-k-out-of-n:F system, *IEEE Trans. Reliab.* 43 (1994) 650-655.

[258] H. Yamamoto, Reliability of a linear connected-(r_1,s_1)-or-(r_2,s_2)-or-...-or-(r_k,s_k)-out-of-(m,n):F lattice system, *Microelectron. Reliab.* 36 (1996) 151-168.

[259] H. Yamamoto and M. Miyakawa, Reliability of a linear connected-(r,s)-out-of-(m,n):F lattice system, *IEEE Trans. Reliab.* 44 (1995) 333-336.

[260] H. Yamamoto and M. Miyakawa, Reliability of a circular connected-(r,s)-out-of-(m,n):F lattice system, *J. Oper. Research Soc. Japan* 39 (1996) 389-406.

[261] Y. C. Yao, On Optimal consecutive-k-out-of-n:F systems subject to a fixed product of failure probabilities, *Prob. Eng. Inform. Sciences* 3 (1989) 165-173.

[262] R. S. Zakaria, H. A. David and W. Kuo, The nonmonotonicity of component importance measures in linear consecutive-k-out-of-n systems, *IIE Trans.* 24 (1992) 147-154.

[263] Y. L. Zhang and Y. Lam, Reliability of consecutive-k-out-of-n:G repairable system, *Inter. J. Syst. Sci.* 29 (1998) 1375-1379.

[264] W. Zhang, C. Miller and W. Kuo, Application and analysis for a consecutive-k-out-of-n:G structure, *Reliab. Eng. Syst. Safety* 33 (1991) 189-197.

[265] Y. L. Zhang and T. P. Wang, Repairable consecutive-2-out-of-n:F system, *Microelecton. Reliab.* 36 (1996) 605-608.

[266] M. J. Zuo, Reliability and component importance of a consecutive-k-out-of-n system, *Microelecton. Reliab.* 33 (1993) 243-258.

[267] M. J. Zuo, Reliability of linear and circular consecutively-connected systems, *IEEE Trans. Reliab.* 42 (1993) 484-487.

[268] M. J. Zuo, Reliability and design of 2-dimensional consecutive-k-out-of-n systems, *IEEE Trans. Reliab.* 42 (1993) 488-490.

[269] M. Zuo, S. Chiovelli and J. Huang, Reliability evaluation of furnace systems, *Reliab. Eng. Syst. Safety* 65 (1999) 283-287.

[270] M. Zuo and W. Kuo, Design and performance analysis of consecutive-k-out-of-n structure, *Nav. Research Log.* 37 (1990) 203-230.

[271] M. J. Zuo and M. Liang, Reliability of multistate consecutively-connected systems, *Reliab. Eng. Syst. Safety* 44 (1994) 173-176.

[272] M. J. Zuo and Y. Wu, Reliability evaluation of a furnace system using the k-out-of-n and consecutive k-out-of-n reliability models, *Proc. IEEE Intern. Conf. on Systems, Man and Cybernetics. Inform. Intell. and Systems* (1996) 3119-3123.

Index

Network Theory and Applications

1. P.-J. Wan: *Multichannel Optical Networks*. 2000 ISBN 0-7923-5776-0
2. B. Fortz: *Design of Survivable Networks with Bounded Rings*. 2000 ISBN 0-7923-6414-7
3. B. Liu and H.J. Lai: *Matrices in Combinatorics and Graph Theory*. 2000 ISBN 0-7923-6469-4
4. G.J. Chang, L. Cui and F.K. Hwang: *Reliabilities of Consecutive-k Systems*. 2000 ISBN 0-7923-6661-1

Kluwer Academic Publishers – Dordrecht / Boston / London

MIX
Papier aus verantwortungsvollen Quellen
Paper from responsible sources
FSC® C105338

If you have any concerns about our products,
you can contact us on
ProductSafety@springernature.com

In case Publisher is established outside the EU,
the EU authorized representative is:
Springer Nature Customer Service Center GmbH
Europaplatz 3, 69115 Heidelberg, Germany

Printed by Libri Plureos GmbH
in Hamburg, Germany